The Shanghai Maths Project

For the English National Curriculum

Practice Book 5A

Series Editor: Professor Lianghuo Fan

UK Curriculum Consultant: Paul Broadbent

Collins

William Collins' dream of knowledge for all began with the publication of his first book in 1819.

A self-educated mill worker, he not only enriched millions of lives, but also founded a flourishing publishing house. Today, staying true to this spirit, Collins books are packed with inspiration, innovation and practical expertise. They place you at the centre of a world of possibility and give you exactly what you need to explore it.

Collins. Freedom to teach.

Published by Collins
An imprint of HarperCollins*Publishers*
The News Building
1 London Bridge Street
London
SE1 9GF

Browse the complete Collins catalogue at
www.collins.co.uk

British Library Cataloguing in Publication Data

A catalogue record for this publication is available from the British Library.

Series Editor: Professor Lianghuo Fan
UK Curriculum Consultant: Paul Broadbent
Publishing Manager: Fiona McGlade and Lizzie Catford
In-house Editor: Mike Appleton
In-house Editorial Assistant: August Stevens
Project Manager: Karen Williams
Copy Editors: Tanya Solomons and Karen Williams
Proofreader: Gerard Delaney
Cover design: Kevin Robbins and East China Normal University Press Ltd.
Cover artwork: Daniela Geremia
Internal design: 2Hoots Publishing Services Ltd
Typesetting: Ken Vail Graphic Design Ltd
Illustrations: QBS
Production: Sarah Burke

Printed and bound by Ashford Colour Press Ltd.

This book is produced from independently certified FSC™ paper to ensure responsible forest management.

For more information visit: www.harpercollins.co.uk/green

The Shanghai Maths Project (for the English National Curriculum) is a collaborative effort between HarperCollins, East China Normal University Press Ltd. and Professor Lianghuo Fan and his team. Based on the latest edition of the award-winning series of learning resource books, *One Lesson, One Exercise*, by East China Normal University Press Ltd. in Chinese, the series of Practice Books is published by HarperCollins after adaptation following the English National Curriculum.

Practice Book Year 5A has been translated and developed by Professor Lianghuo Fan with the assistance of Ellen Chen, Ming Ni, Huiping Xu and Dr Jane Hui-Chuan Li, with Paul Broadbent as UK Curriculum Consultant.

HarperCollins Publishers
Macken House, 39/40 Mayor Street Upper,
Dublin 1, D01 C9W8, Ireland.

Contents

Chapter 1 Revising and improving

1.1 Multiplication and division

Learning objective Multiply and divide numbers with up to 4 digits

Basic questions

1 Use the column method to calculate. Check the answers to the questions marked with *.

(a) 135 × 8 =

(b) 25 × 47 =

(c) 29 × 508 =

  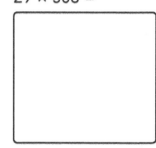

(d) 340 × 890 =

(e) *958 ÷ 9 =

(f) *6500 ÷ 50 =

 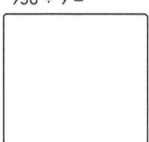 

2 Work these out step by step. (Calculate smartly when possible.)

(a) 32 × 111 − 2058

(b) 76 × 29 + 76 × 21

(c) 108 × 52 − 43 × 71

(d) (1405 − 932) × 33

(e) 680 ÷ (122 − 102)

(f) 238 × (56 + 32)

3 Fill in the spaces to make each statement correct.

(a) The product of the greatest 2-digit number and the smallest 3-digit number is [ ].

(b) The product of 780 × 50 is a [ ]-digit number with [ ] zero(s) at the end.

(c) In □ ÷ 9 = 12 r △, the greatest number that △ could be is [ ]. When △ is that value, □ is [ ].

(d) □70 is a 3-digit number. In the calculation □70 ÷ 50, the quotient is a 1-digit number when the number in the □ is _____. When the number in the □ is _____, the quotient is a 2-digit number.

4 (a) If you use all four of the digits 2, 3, 4 and 5 once each to form all the possible multiplications of two 2-digit numbers, which gives the greatest product and which gives the smallest product? Write them down and then calculate the products.

(b) Without calculating, say which product is greater: 12 345 × 67 890 or 12 346 × 67 889. Explain your reasoning.

1.2 Addition and subtraction of fractions

 Learning objective Add and subtract fractions with the same denominator

 Basic questions

1 Use fractions to represent the shaded part in each of the following figures.

 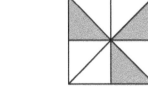 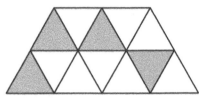

        or

2 Use fractions to represent the shaded part in each of the following figures.

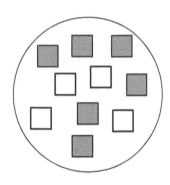 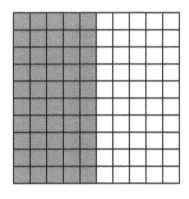

   or   or   or

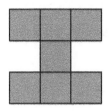

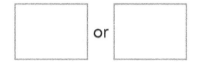

	or	

3 Calculate.

(a) $\frac{1}{5} + \frac{3}{5} =$

(b) $\frac{2}{9} + \frac{3}{9} =$

(c) $\frac{17}{53} + \frac{22}{53} =$

(d) $\frac{56}{143} - \frac{36}{143} =$

(e) $\frac{40}{87} - \frac{13}{87} =$

(f) $\frac{117}{800} + \frac{34}{800} =$

(g) $\frac{8}{19} - \frac{4}{19} + \frac{1}{19} =$

(h) $\frac{52}{111} - \frac{51}{111} + \frac{50}{111} - \frac{49}{111} =$

4 There are 30 pupils in a class. $\frac{3}{5}$ of the pupils are girls.

(a) What fraction of the pupils are boys?

(b) What is the difference between the number of girls and the number of boys? First write your answer as a fraction of the total number of pupils in the class, and then write the answer identifying the number of pupils.

5 A teacher brought 50 books to his class. $\frac{1}{10}$ of them are fiction books, $\frac{3}{10}$ are science books, $\frac{2}{5}$ are storybooks, and the others are mathematics books.

(a) What fraction of the books are mathematics books? ☐

(b) Which type of book were there most of? Which type of book were there least of? Write the four types of books in order, starting from the least.

Challenge and extension questions

6 Complete each calculation with a suitable fraction.

(a) $\frac{5}{16} +$ ☐ $= \frac{1}{2}$

(b) $\frac{79}{100} -$ ☐ $= \frac{3}{10}$

(c) $\frac{19}{22} -$ ☐ $= \frac{3}{22}$

(d) $\frac{37}{100} +$ ☐ $= \frac{1}{2}$

7 A storybook has 96 pages. Joshua plans to read through the whole book in 3 weeks. If he reads $\frac{1}{8}$ of the book in the first week, and $\frac{3}{8}$ of the book in each of the remaining two weeks, can he complete his plan? If not, how many pages will be left? (Hint: Use addition or subtraction of fractions to find your answer.)

1.3 Decimals (1)

Learning objective Read and write decimals and know their fraction equivalents

Basic questions

1 Think carefully and then fill in the missing numbers.

(a) 25.792 consists of [　] tens, [　] ones, [　] tenths,

[　] hundredths and [　] thousandths.

(b) The number that consists of 2 hundreds, 3 tenths and 4 thousandths

is [　].

(c) 15.15 = 10 + [　] + [　] + [　]

(d) 318.79 = 300 + [　] + [　] + 0.7 + [　]

2 Simplify the numbers using the properties of decimals. The first one has been done for you.

(a) 1.50 = 1.5

(b) 0.0110 = [　]

(c) 6.0600 = [　]

(d) 120.000 = [　]

(e) 80.030 = [　]

(f) 16.200 = [　]

3 Rewrite each of the following numbers as a decimal with 3 decimal places without changing its value. The first one has been done for you.

(a) 1.2 = 1.200

(b) 0.56 = [　]

(c) 3 = [　]

(d) 10.2 = [　]

(e) 50.1 = [　]

(f) 120.55 = [　]

4 Write the following fractions as decimals.

(a) $\frac{1}{10} = $ [   ]

(b) $\frac{1}{5} = $ [   ]

(c) $\frac{1}{4} = $ [   ]

(d) $\frac{7}{10} = $ [   ]

(e) $\frac{29}{100} = $ [   ]

(f) $\frac{237}{1000} = $ [   ]

5 Convert these measures using a suitable decimal number.
(Note: £1 = 100p and 1 m = 100 cm.)

(a) 1p = [   ] pounds

(b) 15p = [   ] pounds

(c) 220p = [   ] pounds

(d) £5 and 5p = [   ] pounds

(e) 5 cm = [   ] m

(f) 10 cm = [   ] m

(g) 550 cm = [   ] m

(h) 3 m 18 cm = [   ] m

6 Round the following decimals to their nearest whole numbers.

0.09   0.9   59.3   219.5

[   ]   [   ]   [   ]   [   ]

Challenge and extension question

7 A decimal number with 2 decimal places has the following features:
(1) The value of the whole number part is 5 greater than the greatest 1-digit number.
(2) The sum of the digits in the tenths and hundredths places is 16.
What could the number be?

1.4 Decimals (2)

Learning objective Read, write, order and compare decimals

Basic questions

1 Work on the number line.

(a) Mark these numbers on the number line: 0.5, 10, 5.5, 19.5, 8.5, 14.5, 16.5. (Hint: Use the space both above and below the number line.)

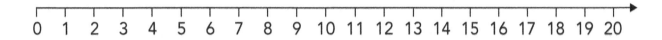

0 1 2 3 4 5 6 7 8 9 10 11 12 13 14 15 16 17 18 19 20

(b) (i) Which one of the numbers in Question (a) is the greatest?

(ii) Which one is the smallest?

(c) Write the numbers in order, starting with the smallest.

_____

2 Write the following decimals as fractions.

(a) 0.1 =

(b) 0.01 =

(c) 0.23 =

(d) 0.2 =

(e) 0.5 =

(f) 0.99 =

3 Fill in the ◯ with >, < or =.

(a) 2.05 ◯ 1.99

(b) 0.75 ◯ $\frac{3}{4}$

(c) 32.35 ◯ 31.78

(d) 0.909 ◯ 0.901

(e) $\frac{1}{10}$ ◯ 0.08

(f) 50.1 ◯ 49.9

4 Multiple choice questions. (For each question, choose the correct answer and write the letter in the box.)

(a) There are ☐ thousandths in 0.1.

    A. 1   **B.** 10   **C.** 100   **D.** 1000

(b) There are ☐ decimals with 1 decimal place greater than 0 but less than 1.

    A. 2   **B.** 5   **C.** 9   **D.** 10

(c) The smallest decimal number with 1 decimal place is ☐.

    A. 0   **B.** 0.1   **C.** 0.01   **D.** 1

(d) The smallest decimal number with 2 decimal places is ☐.

    A. 0.99   **B.** 0.10   **C.** 0.01   **D.** 0.50

5 Comparing decimals.

(a) Put 0.01, 0.1, 0.001 and 0.90 in order, starting from the greatest.

☐ > ☐ > ☐ > ☐

(b) Put the following measures in order, from the shortest to the longest.

1 km   10 m   0.91 km   1.1 km   1.50 km

Challenge and extension question

6 Use 0, 1, 2, 3 and the decimal point to write the decimal numbers described below.

(a) All the decimal numbers less than 1 with 2 decimal places.

_____

_____

(b) All the decimal numbers greater than 2 with 3 decimal places and with a 1 in the tenths place.

_____

_____

(c) All the decimal numbers between 0 and 3 with a 2 in the hundredths place.

_____

_____

1.5 Mathematics plaza – Which area is larger?

Learning objective Calculate and investigate the area and perimeter of squares and rectangles

Basic questions

1 Write your answers in the spaces.

(a) The side length of a square is 1 cm. What is the area of the square?

_____

(b) If the side length of the square is doubled, what is the area of the square?

_____

(c) If the side length of the square is tripled, what is the area of the square?

_____

(d) What is the area of the square if its side length is increased to 4 times its original length?

_____

And if it is increased to 5 times its original length?

_____

(e) What pattern can you see?

_____

_____

2 (a) On the 1 cm square grid paper below, draw all the different squares and rectangles that have a perimeter of 20 cm.

(b) Find the area of each shape. Write your working and answers in the boxes below.

(c) Complete the statements. (Hint: Choose 'the same' or 'different' for the first two answers.)

The perimeters of the shapes drawn in Question (a) are

all _____, but the areas are _____. When

the length and width are _____, the area is greatest.

3 Evie wants to make a rectangular (not including square) window with a perimeter of 140 cm for her lizard's home. How can she design the length and width of the window in order to get the maximum possible sunlight through the window? (Note: Use whole centimetres for the length and width and make a table to investigate the possible dimensions.)

Length (cm)			
Width (cm)			
Area (cm²)			

4 Mr Wood wants to build a rectangular sheep pen. One side of the sheep pen will be an old wall, as shown in the figure. Mr Wood has enough materials to build a new wall 32 m long. Help Mr Wood to design the sheep pen. How long and wide should it be in order to have the maximum possible area? What is the maximum area? (Note: Use whole metres for the length and width and make a table to investigate the possible dimensions.)

Old wall

Sheep pen

5 After the length of a rectangular field is increased by 3 m and the width is increased by 5 m, it becomes a square and its area has increased by 153 m². What was the area of the original rectangular field? (Hint: First draw a diagram to investigate.)

Chapter 1 test

1 Calculate mentally and then write the answers.

(a) 780 + 33 = ▢

(b) 4800 ÷ 60 = ▢

(c) 1200 − 276 = ▢

(d) 104 × 4 = ▢

(e) 20 × 35 = ▢

(f) 98 ÷ 7 = ▢

(g) 60 + 40 ÷ 5 = ▢

(h) (48 − 33) ÷ 3 = ▢

2 Use the column method to calculate. Check the answer to the question marked with *.

(a) 82 × 74 =

(b) 860 × 305 =

(c) *3400 ÷ 40 =

3 Work these out step by step. (Calculate smartly when possible.)

(a) 125 × 32 × 25

(b) 61 × 89 + 61 × 11

(c) $385 \div (215 - 160)$

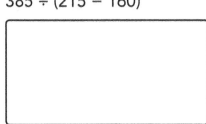

(d) $144 \times 59 \div 16$

4 Add or subtract these fractions.

(a) $\frac{1}{7} + \frac{4}{7} = $ [ ]

(b) $\frac{2}{15} + \frac{11}{15} = $ [ ]

(c) $\frac{19}{29} - \frac{13}{29} = $ [ ]

(d) $\frac{184}{365} - \frac{28}{365} = $ [ ]

(e) $\frac{18}{49} + \frac{15}{49} - \frac{30}{49} = $ [ ]

(f) $\frac{87}{100} - \left(\frac{33}{100} + \frac{11}{100}\right) = $ [ ]

5 Write the following decimals as fractions.

(a) $0.3 = $ [ ]

(b) $0.03 = $ [ ]

(c) $0.33 = $ [ ]

(d) $0.71 = $ [ ]

(e) $0.7 = $ [ ]

(f) $0.87 = $ [ ]

6 Fill in the missing numbers to make each statement correct.

(a) 23.353 consists of [ ] tens, [ ] ones, [ ] tenths and [ ] hundredths and [ ] thousandths.

(b) $158.39 = 100 + $ [ ] $ + $ [ ] $ + 0.3 + $ [ ] .

(c) Simplify these numbers using the properties of decimals.

$1.050 = $ [ ] , $0.60 = $ [ ] ,

and $365.00 = $ [ ] .

(d) There are [ ] zeros at the end of the product of 150×60.

(e) If the quotient of $8\overline{)\square 96}$ is a 3-digit number, the smallest possible number in the $\square$ is [   ]. In this case, the quotient is [   ].

If the quotient is a 2-digit number, the greatest possible number in the $\square$ is [   ]. In this case, the quotient is [   ] and the remainder is [   ].

(f) If you divide a 1 m long wire into 100 parts equally, the length of each part is [   ] m or [   ] cm. The length of 11 parts is [   ] m or [   ] cm.

(g) The length of a rectangle is 10 cm. If the length of the rectangle is increased by 2 cm, and its width remains unchanged, the area increases by 16 cm². The perimeter of the original rectangle is [   ] cm.

(h) If the area of a square is 81 m², the perimeter is [   ] m.

(i) 25.49 rounded to the nearest whole number gives [   ].

50.50 rounded to the nearest whole number gives [   ].

7 Use fractions to represent the shaded part in each of the following figures.

[   ] or [   ]   [   ]   [   ] or [   ]

8   Solve these problems.

(a) The weight of a baby bear is 69 kg. This is equal to the weight of 3 monkeys. What is the difference between the weight of a monkey and the weight of a baby bear?

_____

(b) There are 60 balls in a bag. Half of them are black, one third are white and the remaining balls are red. What fraction of the balls are red? How many of them are there?

_____

(c) The perimeter of a rectangular flowerbed is 68 m. The length is 25 m. What is the area of this rectangular flowerbed?

_____

(d) A rope is exactly long enough to enclose a rectangle with a length of 18 m and a width of 10 m. If the same rope is used to enclose a square, what is the area of the square?

_____

(e) Six identical squares with a side length of 3 cm are put together to form a large rectangle. What is the maximum perimeter of the possible rectangles formed?

_____

Chapter 2 Large numbers and measures

2.1 Knowing large numbers (1)

 Learning objective Recognise the place value of the digits in large numbers

 Basic questions

1 Fill in the spaces to make each statement correct.

(a) Counting from the right, a large number can be separated into groups: ones group, thousands group, millions group, with each group containing three places for different values: ones, tens and hundreds. Complete the table below.

Group	Millions			Thousands			Ones		
Place value		Ten millions				Thousands		Tens	

(b) The value of a digit is ▢ times the value of the same digit in the place to its right.

(c) 378 028 867 consists of ▢ millions, ▢ thousands and ▢ ones.

(d) When reading a large number, we start from the left with the largest group. For a number whose largest group is millions, we first read the millions group, then the _____ group and finally the

_____ group. 378 028 867 is read as _____

_____

_____

(e) When writing a large number of five or more digits in numerals, we start from the left with the largest group, and leave a space between each group (counting from right to left). Twenty-one million, one thousand and thirty-six is written as [   ] .

2 Complete the place value chart for each number and fill in the blanks. The first one has been done for you.

(a) 4019

Thousands	Hundreds	Tens	Ones
4	0	1	9

Read as: <u>Four thousand and nineteen</u>

$4019 = $ <u>$4 \times 1000 + 0 \times 100 + 1 \times 10 + 9 \times 1$</u>

$ = $ <u>$4000 + 0 + 10 + 9$</u>

(b) 25 198

Ten thousands	Thousands	Hundreds	Tens	Ones

Read as: _____

_____

$25\,198 = $ _____ + _____ + _____ + _____ + _____

$ = $ _____ + _____ + _____ + _____ + _____

(c) 412 708

Hundred thousands	Ten thousands	Thousands	Hundreds	Tens	Ones

Read as: _____

_____

412 708 = _____ + _____ + _____ + _____ + _____

+ _____

= _____ + _____ + _____ + _____ + _____

+ _____

(d) 2 395 198

Millions	Hundred thousands	Ten thousands	Thousands	Hundreds	Tens	Ones

Read as: _____

_____

2 395 198 = _____ + _____ + _____ + _____ + _____

+ _____ + _____

= _____ + _____ + _____ + _____ + _____

+ _____ + _____

(e) 76 203 000

Ten millions	Millions	Hundred thousands	Ten thousands	Thousands	Hundreds	Tens	Ones

Read as: _____

_____

76 203 000 = _____ + _____ + _____ + _____ + _____

+ _____ + _____ + _____

= _____ + _____ + _____ + _____ + _____

+ _____ + _____

3 Fill in the table. The first one has been done for you.

Number in words	Number in numerals
One million, two hundred and one	1 000 201
One hundred and nineteen thousand and thirty-three	
Seventy million, seventy thousand and seven	
Nine hundred and sixty-one million, two hundred and seventy-three thousand, nine hundred and twenty-eight	
Five hundred million	

4 True or false? (Put a ✓ for true and a ✗ for false in each box.)

(a) The number consisting of two ten thousands, two one thousands, and two ones is 2 000 020 002.

(b) 50 040 600 is read as fifty thousand and forty thousand and six hundred.

(c) In a 5-digit number, the place with the highest value is the ten thousands place. ☐

(d) Nine hundred and nine thousand and nine is written in numerals as 909 009. ☐

Challenge and extension questions

5 Use four zeros and three sixes to form different 7-digit numbers as indicated.

(a) Four zeros at the end: _____

(b) Three zeros at the end: _____

(c) Two zeros at the end: _____

6 Use the six digits 3, 0, 1, 5, 7 and 9 to form the greatest and smallest possible 6-digit numbers. What is the difference between these two numbers? (Write a number sentence to show your answer.)

_____

_____

_____

2.2 Knowing large numbers (2)

Learning objective Compare and order large numbers

Basic questions

1 Fill in the spaces to make each statement correct.

(a) In a 9-digit number, the place with the highest value is in the

_____ place.

(b) Counting from right to left, the fifth place is the

_____ place, the place to its right is the

_____ place, and the place to the left of

the hundred thousands place is the _____
place.

(c) 2 003 500 709 contains ⬜ billions, ⬜ millions, ⬜

thousands and ⬜ ones.

(d) In the number 10 506 000 090, the 1 is in the

_____ place, standing for

_____ ; the 5 is in the

_____ place, standing for

_____ ; the 6 is in the

_____ place, standing for

_____ ; and the 9 is in the

_____ place, standing for

_____ .

(e) 913 004 consists of ⬚ thousands and ⬚ ones. It can

also be said to consist of ⬚ ones.

(f) 4000 can be seen as ⬚ ones, or ⬚ thousands, or

⬚ hundreds.

(g) The smallest 7-digit number is ⬚ , the smallest 4-digit

number is ⬚ , and the smallest 7-digit number is ⬚

times the smallest 4-digit number.

2 Multiple choice questions. (For each question, choose the correct answer and write the letter in the box.)

(a) Two hundred and two thousand three hundred is written as ⬚ .

  A. 2 002 300   **B.** 202 000 300 **C.** 202 300   **D.** 2 020 300

(b) In the number 6 104 510, the 4 is in the ⬚ place.

  A. hundreds   **B.** thousands

  C. ten thousands   **D.** hundred thousands

(c) One billion is equal to ⬚ millions.

  A. 10   **B.** 100   **C.** 1000   **D.** 10 000

3 Write >, < or = in each ◯.

(a) 34 796 ◯ 43 796

(b) 100 001 ◯ 99 999

(c) 900 100 ◯ 1 000 000

(d) Thirty million, three hundred thousand ◯ 30 300 000

4 Write the numbers that are greater than 69 997 and less than 70 003, and then represent them on the number line.

(a) The numbers: [＿＿＿] , [＿＿＿] , [＿＿＿] ,

[＿＿＿] , [＿＿＿]

(b) The number line:

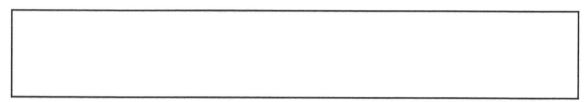

69 997

5 Subtract 23 from the smallest possible 5-digit number and add the resulting number to the largest possible 3-digit number. Show the calculations.

[＿＿＿＿＿＿＿＿＿＿＿＿＿＿＿＿＿＿＿＿＿＿＿]

6 Calculate mentally or use the column method to add or subtract large numbers. Check the answers to the questions marked with *.

(a) 608 000 + 200 000

(b) 236 549 – 36 000

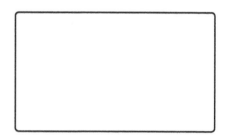

(c) *720 055 – 22 000

(d) 32 495 – 30 001

(e) 225 329 + 154 019

(f) *5 800 000 − 712 012

Challenge and extension questions

7 Put the numbers 60 500, 50 600, 500 006, 56 000 and 65 000 in order, first from the smallest to the greatest, and then from the greatest to the smallest.

8 Find the populations of England, Scotland, Wales and Northern Ireland, and then add them up to find the total population of the United Kingdom. (State the source of your data.)

2.3 Knowing large numbers (3)

Learning objective Read, write, order and compare large numbers

Basic questions

1 Fill in the spaces to make each statement correct.

(a) 90 301 000 is an ☐-digit number. The digits 9, 3 and 1 are in the

_____ place, the _____

place and the _____ place, respectively.

It is read as: _____

_____.

(b) The number consisting of 60 millions, 54 thousands and 3 ones is

written ☐ in numerals.

(c) Six hundred and sixty million, sixty thousand and six is written as

☐ in numerals. It is a ☐-digit number.

(d) The digit in both the ten millions place and the thousands place is 6,
the digit in the tens place is 8, and the digit in all other places is 0. The

number is ☐, read as _____

_____.

(e) Eight hundred thousand has ☐ ten thousands, and four hundred

and thirty million has ☐ ten millions.

(f) Adding ☐ zeros after fifty-seven, the resulting number is read as
five hundred and seventy million.

2 Read the numbers underlined and write them in words below.

(a) According to the national statistics, the number of pupils in schools in England in 2015 was <u>8 271 270</u>.

_____

_____

(b) According to the Guinness World Records, the largest pizza weighed <u>23 250</u> kg.

_____

_____

(c) The Coral Sea is in the South Pacific Ocean, off the north-east coast of Australia. It contains the world's largest reef system and has an area of <u>4 791 000</u> square kilometres.

_____

_____

(d) The distance between Pluto and the Sun is about <u>5 900 000 000</u> km.

_____

_____

3 True or false? (Put a ✓ for true and a ✗ for false in each box.)

(a) The greatest 7-digit number is 9 000 000.

(b) The number consisting of 70 thousands and 500 ones is 700 500.

(c) A large number plus a large number is also a large number.

(d) A large number minus a large number is also a large number.

4 Draw a line to match each pair of the same number.

(a) four hundred and five thousand and eight   **A.** 400 050 008

(b) forty million, fifty thousand and eight   **B.** 40 050 008

(c) four hundred million, fifty thousand and eight   **C.** 40 000 080

(d) four billion, fifty million, eight hundred   **D.** 405 008

(e) forty million and eighty   **E.** 4 050 000 800

(f) Now write all the numbers in order, from the greatest to the least:

Challenge and extension questions

5 You are given two numbers: 5976 and 1432. Subtract 4 from the first number and add 4 to the second number. Repeat. How many times do you have to do this before they become equal?

6 Joe and Alan are playing number games. Joe says, 'I am thinking of a 6-digit number, which is greater than the greatest 5-digit number. The digit in its highest value place is 1, and the digit in all the other places is 0.' Alan got the correct answer immediately. What is the number?

7 Remove four digits from the 10-digit number 5 704 590 212 so that, without changing the order, the 6-digit number made up of the remaining six digits has the greatest value. What is the 6-digit number? Repeat the activity, but this time remove four digits to leave the 6-digit number with the smallest value. What is the number?

2.4 Rounding of large numbers (1)

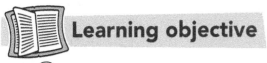

Learning objective Round numbers to the nearest 100, 1000 and 10000

Basic questions

1 Circle the whole hundreds number that each number on the upper part of the number line is nearest to.

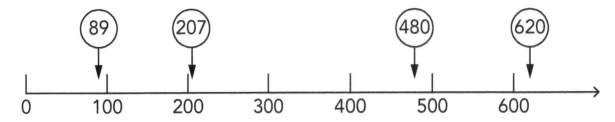

2 Write the whole ten thousands numbers that come before and after a, b, c, d and e. Put a ✓ against the ten thousands number that each number is nearest to.

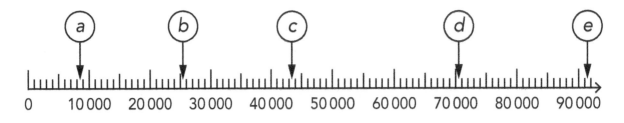

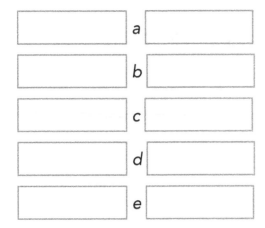

3 Write the whole ten thousands numbers before and after each of the following numbers. Put a ✓ next to its nearest ten thousands number.

	32 108	
	105 213	
	971 234	
	120 087	
	6 401 239	
	396 042	

4 Round each of the following numbers to the nearest ten thousand.

(a) 10 999 ≈

(b) 56 089 000 ≈

(c) 443 219 ≈

(d) 1 096 789 ≈

(e) 9167 ≈

(f) 9 950 123 ≈

5 Approximate these measures.

(a) £81 023 ≈ _____ thousand pounds

(b) 119 412 m ≈ _____ thousand metres

(c) 2 095 802 m ≈ _____ million metres

(d) £999 999 ≈ _____ million pounds

6 A wholesale fruit market has 136090 kg of apples, 59400 kg of bananas, 70020 kg of oranges and 1064999 kg of pears.

(a) Read and write these numbers in words:

136090: _____

_____

59400: _____

_____

70020: _____

_____

1064999: _____

_____

(b) Round each number to the nearest ten thousand.

136090 ≈ [_____]   59400 ≈ [_____]

70020 ≈ [_____]   1064999 ≈ [_____]

(c) Write the numbers in order, starting from the greatest.

_____

_____

Challenge and extension questions

7 Complete each statement (rounding to the nearest whole number).

(a) Use 3 zeros and 4 sevens to make a number. The greatest number is

[   ] , which is read as

_____

_____ .

Rounding it to the nearest ten thousand, the result is

[   ] .

(b) To make 56 ▨ 13 589 ≈ 56 million true, the digit in the ▨ could be

[   ] . The greatest possible digit in the ▨ is [   ] .

(c) To make 1 09 ▨ 028 ≈ 1 million and 100 thousands, ▨ could be [   ] .

The smallest possible digit in the ▨ is [   ] .

(d) 10 056 217 101 ≈ [   ] billion.

8 Use 2 sevens and 5 zeros to make two different 7-digit numbers with the smallest possible difference. Write both numbers.

2.5 Rounding of large numbers (2)

Learning objective Round numbers to the nearest 1000, 10 000, 100 000 and 1 000 000

Basic questions

1 Write the whole hundred thousands numbers that come before and after a, b, c, d and e. Put a ✓ next to the hundred thousands number that each number is nearest to.

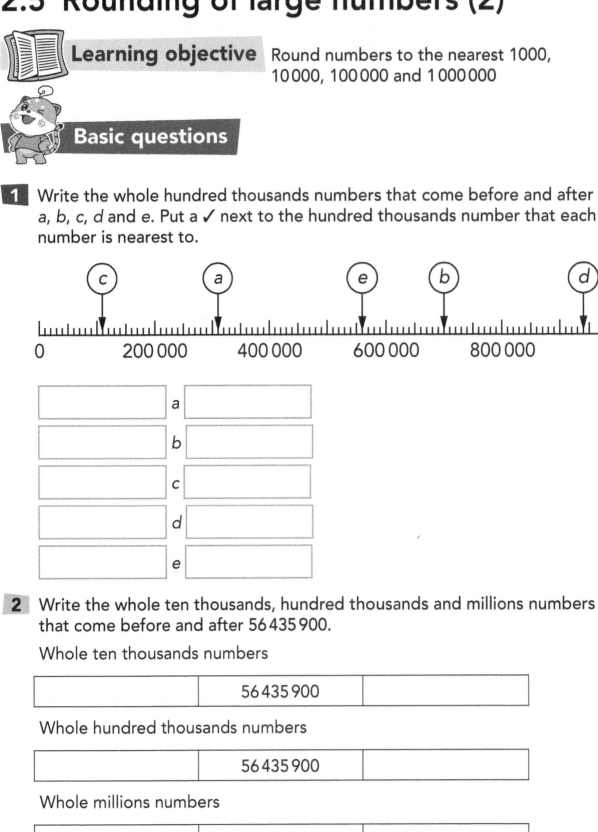

	a	
	b	
	c	
	d	
	e	

2 Write the whole ten thousands, hundred thousands and millions numbers that come before and after 56 435 900.

Whole ten thousands numbers

	56 435 900	

Whole hundred thousands numbers

	56 435 900	

Whole millions numbers

	56 435 900	

3 Fill in the boxes to make each statement correct.

When we round a number, we first decide which digit is the last digit to keep. If the next digit to its right is ☐ or more, we increase it by 1 and all the digits to its right change to zero (known as rounding up). If the next digit is less than ☐, we leave it the same and all the digits to its right change to zero (known as rounding down).

4 Round the following numbers to the nearest 1000, 10 000, 100 000 and 1 000 000.

	5 375 021	9 988 522	1 240 641	1 000 234
Nearest 1000				
Nearest 10 000				
Nearest 100 000				
Nearest 1 000 000				

5 A 5-digit number, after being rounded to the nearest 10 000, is 90 000. What is the greatest possible value of this number? What is its smallest possible value?

Challenge and extension question

6 A number consists of 6 hundred thousands, 4 thousands, 5 hundreds and 3 ones. What number is it? When it is rounded to the nearest hundred thousand, what is the resulting number? What is the smallest number it can be added to so that when the sum is rounded the result is 610 thousand?

2.6 Converting kilograms and grams

Learning objective Convert between kilograms and grams and solve measures problems

Basic questions

1 The relationship between kilograms and grams is:

1 kilogram = [ ] grams

2 Write a suitable unit in each space: kilograms (kg) or grams (g).

(a) The mass of a puppy is about 2 _____.

(b) The mass of a bag of crisps is about 250 _____.

(c) The mass of a rhinoceros is about 2000 _____.

(d) A pear weighs about 100 _____.

(e) Ben weighs about 30 _____.

(f) A rubber weighs about 6 _____.

3 Fill in the boxes.

(a) 307 kg = [ ] g

(b) 17 kg = [ ] g

(c) 6 000 000 g = [ ] kg

(d) 1025 g + 71 kg = [ ] g

(e) 43 000 kg + 4000 g = [ ] kg

(f) 67 kg − 20 202 g = [ ] g

4 Write >, < or = in each ◯.

(a) 5 kg ◯ 500 g

(b) 6000 g ◯ 6 kg

(c) 7800 kg ◯ 8 000 000 g

(d) 90 000 kg ◯ 9 000 000 g

(e) 13 000 kg ◯ 1000 kg + 200 kg

(f) 4000 kg ◯ 7000 g

5 Multiple choice questions. (For each question, choose the correct answer and write the letter in the box.)

(a) A truck is loaded with 3 machine tools and 450 kg of accessories. Each machine tool weighs 800 kg. The truck is loaded with the total mass of ☐.

A. 2050 kg   **B.** 2850 kg   **C.** 2150 kg   **D.** 1250 g

(b) There is 482 000 kg of sand in a builders' yard. If a dumper truck can be loaded with 4000 kg of sand, how many trucks will be needed to transport all the sand in one go? ☐

A. 120 trucks   **B.** 121 trucks   **C.** 122 trucks   **D.** 123 trucks

(c) 30 kg 900 g = ☐ g

A. 30 900   **B.** 903   **C.** 1200   **D.** 3900

6 Solve these problems.

(a) Tom's mother bought 4 bottles of sunflower oil. Each bottle weighs 2500 g. Find the total mass of the 4 bottles, first in grams and then in kilograms.

_____

(b) 8 electrical machines are loaded onto a truck, which has a loading capacity of 5000 kg. The mass of each machine is 620 kg. Does the total mass of all the machines exceed the loading capacity of the truck?

_____

(c) Martha's uncle harvested 5000 kg of pears in his orchard last year into boxes. If each box can be filled with 25 kg of pears, how many boxes can be filled with all the pears?

Challenge and extension questions

7 True or false? (Put a ✓ for true and a ✗ for false in each box.)

(a) 1000 g of cotton is lighter than 1 kg of iron.

(b) There are two lots of 10 000 grams in 20 kilograms.

(c) 1 kg is 30 g heavier than 70 g.

8 The mass of a box of apples was 46 kg inclusive of the mass of the box. After selling half the apples, the mass was 26 kg. What was the mass of the box?

2.7 Litres and millilitres (1)

Learning objective Convert between litres and millilitres and solve measures problems

Basic questions

1 Fill in the spaces to make each statement correct.

(a) The amount of liquid is often expressed in millilitres and litres. When measuring a small amount of liquid, we usually use millilitres as the unit

of measure. When measuring a _____ amount of liquid,

we usually use _____ as the unit of measure.

(b) 1 millilitre can be written as 1 ml. 1 litre can be written as 1 l or 1 L.

1 l = [_____] ml.

2 Which measuring cups of water can be added together to fill up a 1 litre bottle?

Write the number sentence. _____

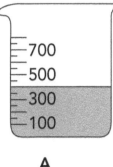

A

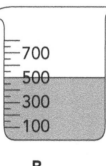

B

C

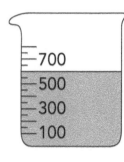

D

3 Write a suitable unit in each space: litres or millilitres.

(a) A tank of petrol: 30 _____

(b) A bottle of cooking oil: 2 _____

(c) A tub of yogurt: 250 _____

(d) A bottle of eye drops: 5 _____

(e) A cup of water: 200 _____

(f) A bottle of cola: 600 _____

(g) A bottle of liquid medicine: 10 _____

4 Convert these measures.

(a) 2l = _____ ml

(b) 800l = _____ ml

(c) 82 000 ml = _____ litres

(d) 50l = _____ ml

(e) 21 600 ml = _____ l

(f) 30 003 ml = _____ l _____ ml

5 Solve these problems.

(a) Samia bought a 2 litre bottle of orange juice. After pouring 400 ml into a cup, how many millilitres of orange juice were left?

_____

(b) A 2 litre bottle of orange juice was shared equally among 8 children. How many millilitres of orange juice did each child get?

_____

Challenge and extension questions

6 Multiple choice questions. (For each question, choose the correct answer and write the letter in the box.)

(a) $3\,l\,30\,ml = \boxed{}\ ml$

    A. 3300   **B.** 3030   **C.** 3003   **D.** 303

(b) $7000\,ml$ and $3\,l$ is equal to $\boxed{}$.

    A. 7003 ml   **B.** 1000 ml   **C.** 10 ml   **D.** 10 l

(c) 4 bottles of 600 ml of a drink is equal to $\boxed{}$ bottles of 200 ml of the drink.

    A. 3   **B.** 6   **C.** 12   **D.** 18

(d) 1 litre of juice is shared by 16 children in a nursery. Each child gets 50 ml. There are $\boxed{}$ left.

    A. 800 ml   **B.** 20 ml   **C.** 2 l   **D.** 200 ml

7 375 ml of concentrated grape juice was added to 11 litres of water and then equally shared by 25 workers. How many millilitres of grape juice did each worker get?

2.8 Litres and millilitres (2)

Learning objective Convert between litres and millilitres and solve measures problems

Basic questions

1 Draw lines to match the pairs. Use a ruler.

(a) a spoon of medicine   **A.** 1000 ml

(b) a can of soft drink   **B.** 5 l

(c) a bottle of cooking oil   **C.** 330 ml

(d) a bottle of milk   **D.** 18 l

(e) a bucket of water   **E.** 5 ml

2 Fill in the spaces.

(a) When measuring the amount of a liquid such as water or oil,

we can use _____ and _____ as units of measurement.

(b) 1000 ml of water can fill up a water bottle of ⬚ litre(s).

(c) 2 l of beverage can fill up ⬚ cups of 500 ml each.

(d) 13 l = _____ ml   (e) 12 000 ml = _____ litres

(f) 10 000 ml = _____ litres   (g) 51 500 ml = _____ ml

(h) 2 l + 21 000 ml = _____ litres   (i) 41 567 ml = _____ ml

(j) 56 010 ml = _____ l _____ ml   (k) 10 700 ml = _____ l _____ ml

3 Write the following quantities in order.

(a) From the greatest to the least.

| 600 ml | 5900 ml | 7 l | 5970 ml | 70 l |

_____

(b) From the least to the greatest.

| 260 ml | 2060 ml | 2 l | 200 ml | 20 l |

_____

4 Solve these problems.

(a) 16 crates of coconut milk were delivered to a supermarket. Each crate had ten 350 ml cartons of coconut milk. How many litres of coconut milk were delivered?

_____

(b) There are 625 ml of soya milk in 5 packs of the same size. How many millilitres of soya milk are there in 8 packs? How many litres are there?

_____

Challenge and extension question

5 Convert the measures based on the diagrams shown below.

(a)

$1\,l$ = _____ ml

(b)

$\frac{1}{2}\,l$ = _____ ml

(c)

$\frac{1}{4}\,l$ = _____ ml

Three $\frac{1}{4}\,l$ = _____ ml

Two $\frac{3}{4}\,l$ = _____ ml

(d)

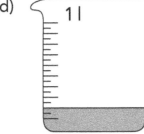

$\frac{1}{5}\,l$ = _____ ml

Two $\frac{1}{5}\,l$ = _____ ml

Three $\frac{2}{5}\,l$ = _____ ml

Chapter 2 test

1 Calculate mentally and then write the answers.

(a) $90 \div 45 \times 2 =$ ☐

(b) $81 \div 3 + 9 =$ ☐

(c) $800 \div 20 - 10 =$ ☐

(d) $1000 - 90 =$ ☐

(e) $38 \div 19 \times 100 =$ ☐

(f) $240 - 15 \times 2 =$ ☐

(g) $280 \times 2 =$ ☐

(h) $4000 \div 100 =$ ☐

(i) $803 - 79 =$ ☐

2 Use the column method to find the answer to each calculation.

(a) 403×16

(b) $8098 - 909$

(c) $16\,018 + 994$

(d) $1600 \div 60$

3 Work these out step by step.

(a) 6129 − (715 + 1129) − 285

(b) 3279 + 480 ÷ 30

(c) 5760 ÷ 30 × 72

(d) 13 320 − 222 × 56

4 Fill in the spaces to make each statement correct.

(a) Counting from the right, the places in a number from the first place to

the _____ place are grouped as ones group. They stand

for ones, _____ and _____. The places from

the fourth place to the _____ place are grouped

as thousands group. They stands for thousands, _____

and _____. The places from the seventh place to the

_____ place are grouped as millions group. They stand

for _____, _____ and _____.

(b) 9 025 000 is read as _____

_____.

It has ☐ millions and ☐ thousands.

(c) 70 700 000 is read as _____

_____.

It has [   ] thousands.

(d) 3 080 604 is read as _____

_____.

It has [   ] thousands and [   ] ones.

(e) A 6-digit number has a 3 in the highest value place and another 3 in the lowest value place. All the other digits are zeros. The number is

[   ]. It is read as _____

_____.

(f) Four million, thirty-four thousand and twenty is written in numerals as

[   ].

(g) 1 023 757 ≈ _____ (to the nearest ten thousand)

(h) 659 313 664 ≈ _____ (to the nearest ten million)

(i) 90 000 m = _____ km

(j) 20 l + 350 ml = _____ ml

(k) [   ] kg − 3000 kg = 5300 kg

(l) Put the following measures in order, starting with the least: 3 litres 400 millilitres, 7080 millilitres, 6 litres, and 10 litres 7 millilitres.

_____

_____

(m) $5\,000\,000\,g =$ _____ kg

(n) ⬚ g = 300 kg

5 Multiple choice questions. (For each question, choose the correct answer and write the letter in the box.)

(a) The numbers that come before and after 20 000 are ⬚.

   A. 19 999 and 1999   **B.** 19 000 and 20 001   **C.** 19 999 and 20 001

(b) If 39 ▨ 270 ≈ 400 000 to the nearest 10 000, the numbers that can be

   filled in the ▨ are ⬚.

   A. 9, 8, 7, 6 or 5   **B.** 4 or 5   **C.** 0, 1, 2, 3 or 4

(c) £9 000 550 rounded to the nearest 10 000 is ⬚.

   A. 9 million pounds   **B.** 9 million and 1 thousand pounds

   C. 10 million pounds

(d) A rhino weighs about 2000 ⬚ , a bag of sugar weighs about

   1000 ⬚ , a cat weighs about 4 ⬚ and a bucket of water

   weighs about 19 ⬚ .

   A. grams   **B.** kilograms   **C.** litres

(e) In 805 794, the value that the digit 8 stands for is ⬚ times the value that the digit 4 stands for.

   A. 20   **B.** 20 000   **C.** 200 000

(f) Adding ⬚ zeros to the right of 37, makes 37 million.

   A. 4   **B.** 5   **C.** 6

6 Write the number sentences and then calculate.

(a) Subtracting a number from 330, the result is 98. Find the number.

(b) In a division, the quotient is 6 and the sum of the dividend and the divisor is 98. Find the dividend and the divisor.

(c) What is 4 times the difference between 78 and 31?

(d) The product of 23 and 105 is divided by 30. What are the quotient and remainder?

7 Solve these problems.

(a) How many 300 ml bottles can an 18 l bucket of water fill up?

(b) A restaurant bought 60 bags of rice and 40 bags of flour. Bags of both rice and flour weigh 50 kg each. What is the total weight of these bags of rice and flour?

(c) 74 000 kg of cement was delivered to a construction site. The mass of sand delivered to the site was 12 000 kg more than 3 times the mass of cement. What is the mass of the sand delivered to the site?

(d) A rectangular footpath is 300 m long and 6 m wide. If 50 cm × 50 cm square slates are used to pave the footpath, how many slates are needed?

(e) Jenna's mother bought a fish tank with a small hole in it. She turned on a tap to pour water at 300 ml per minute into the tank, but at the same time the hole was leaking water at 40 ml per minute from the tank. Five minutes later, Jenna's mother noticed the leakage. At that time, how many millilitres of water were still in the fish tank? How many millilitres of water had leaked out?

3.1 Speed, time and distance (1)

Learning objective Calculate speed using given distance and time

Basic questions

1 True or false? (Put a ✓ for true and a ✗ for false in each box.)

(a) Speed tells how fast a moving object travels. ☐

(b) Distance tells how much time a moving object has spent in motion. ☐

(c) If Mr Lee walks 2 km in two hours, then his speed is 1 km per hour. ☐

(d) If a bird flies at a speed of 2 m per second, then it can fly 120 m per minute. ☐

(e) In a running race, the winner runs at the fastest speed and is the first person to cross the finish line. ☐

2 Fill in the table.

	Speed	Time	Distance
		3 hours (h)	42 km
		5 minutes (min)	4500 m
		20 seconds (s)	400 m

3 Use a suitable unit (km/h, km/min, m/min) to complete each statement.

The pigeon flies at a speed of 65 _____ .	The lion runs at a speed of 1 _____ .
The elephant walks at a speed of 20 _____ .	The puppy runs at a speed of 330 _____ .

4 Read the following carefully and then calculate and compare.

To celebrate a holiday, Mary, Joan, Tom and John planned to meet each other at the gate of a park at 9 o'clock to watch shows in the park. They arrived on time.

Mary left home at a quarter to nine. It is 1500 m from her home to the park.

Her speed was _____ .

Joan left home at 10 to nine. It is 1200 m from her home to the park.

Her speed was _____ .

John left home at 8 minutes to nine. It is 1000 m from his home to the park.

His speed was _____ .

Tom left home at 10 to nine. It is 1300 m from his home to the park.

His speed was _____ .

Which of them walked the fastest? _____ .

5 Solve these problems.

(a) Andrew walked to Jim's house for a party. The distance between their houses is 800 m. Andrew left home at 10 past 10 and arrived at 18 minutes past 10.

Find Andrew's walking speed. _____

(b) The distance between place A and place B is 800 km. A train leaves place A at 5 o'clock and arrives at place B at 10 o'clock.

Find the speed of the train. _____

Challenge and extension question

6 The distance of the London marathon is 42 km and 195 m. Famous British marathon runner Paula Radcliffe is the women's world record holder in the marathon with her time of 2 hours, 15 minutes and 25 seconds.

Her running speed is about _____ m/s.

3.2 Speed, time and distance (2)

Learning objective Solve problems using the relationship between time, distance and speed

Basic questions

1 Fill in the blanks. The first one has been done for you.

Distance = Speed (×) Time

Time = _____ ◯ _____

Speed = _____ ◯ _____

2 Compare and then write the answers.

(a) The distance between the school and the museum is 1 km. Both Bill and Jim walked to the museum from their school. Bill left at 16:00 and arrived at 16:15. Jim left at 16:05 and arrived at 16:18.

Who walked faster? _____

When the distance is the same, we can compare the time.

The less the time, the _____ the speed.

(b) Emmy and Joanne left school and walked home at 15:30. By 15:50 both of them reached their homes. The distance between Emmy's home and the school is 1500 m while the distance between Joanne's home and the school is 1650 m.

Who walked faster? _____

When the time taken is the same, we can compare the distance.

The greater the distance travelled, the _____ the speed.

3 Fill in the table.

Speed	Time	Distance
	6 min	504 m
7 km/h		119 km
118 m/min	8 min	

4 Solve these problems.

(a) It took Emily 5 minutes to walk 400 m while it took Samantha 4 minutes to walk 360 m. Who walked faster?

_____

(b) Alvin and his mother left home at 7 o'clock in the morning. His mother drove to her office at a speed of 700 m/min and Alvin cycled to school at a speed of 16 km/h. Both of them reached their destinations at half past seven. Find out the distance between Alvin's mother's office and their home. How about the distance between Alvin's home and his school?

_____

_____

_____

(c) A bicycle travels at 18 km/h and a motorcycle travels 27 km/h faster than the bicycle. How many kilometres can a motorcycle travel in 8 hours?

_____

_____

_____

(d) Tim left home at 7 o'clock in the morning for school. After walking for 2 minutes he realised he had left his maths homework at home. So he went back to get his homework and arrived at the school at 28 minutes past 7. Given Tim walked at 100 m/min, and it took him 4 minutes to get home and find the homework, what is the distance between his home and the school?

Challenge and extension question

5 Four pupils were having a 100 m running race. Lily ran at 8 m/s, Anna took 13 seconds, Linda took 12 seconds and Mary ran at 7 m/s.

Who ran the fastest among the four pupils? _____

3.3 Dividing 2-digit or 3-digit numbers by a 2-digit number (1)

 Learning objective Use mental and written methods to divide 3-digit numbers by 2-digit numbers

 Basic questions

1 Work these out mentally and then write the answers.

(a) $6 \div 2 =$ ☐   (b) $8 \div 4 =$ ☐   (c) $9 \div 3 =$ ☐   (d) $21 \div 7 =$ ☐

(e) $60 \div 2 =$ ☐   (f) $80 \div 4 =$ ☐   (g) $90 \div 3 =$ ☐   (h) $210 \div 7 =$ ☐

(i) $600 \div 20 =$ ☐   (j) $80 \div 40 =$ ☐   (k) $90 \div 30 =$ ☐   (l) $210 \div 70 =$ ☐

(m) $600 \div 200 =$ ☐   (n) $800 \div 400 =$ ☐   (o) $900 \div 300 =$ ☐   (p) $2100 \div 700 =$ ☐

2 Answer the questions about each calculation and then calculate the answer.

(a) In $99 \div 23 =$

How many twenty-threes are there in 99?

Think: How many _____ are there in 99?

$\overline{)}$ There are _____ _____ in 99. The quotient is _____.

_____ $\times 23 =$ _____.

The remainder _____ is _____ than the divisor.

The quotient is _____.

(b) In 517 ÷ 63 =

How many sixty-threes are there in 517?

Think: How many _____ are there in 517?

There are _____ _____ in 517. The quotient is _____.

_____ × 63 = _____.

The remainder _____ is _____ than the divisor.

The quotient is _____.

3 Use the column method to calculate.

174 ÷ 21 =   195 ÷ 53 =   132 ÷ 22 =

222 ÷ 43 =   347 ÷ 74 =   608 ÷ 86 =

4 These animals are having a 1 km running race. Look at their speeds and answer the questions.

12 m/s   10 m/s   13 m/s

In 1st place is _____

Reason: _____

_____.

In 2nd place is _____

Reason: _____

_____.

In 3rd place is _____

Reason: _____

_____.

Challenge and extension question

5 Which transport to hire?

18 seats for £200 49 seats for £480

182 pupils and teachers in Year 5 are going on a trip to a local science museum.

(a) If they hire 49-seater coaches, at least how many coaches do they need? What is the cost?

_____

(b) If they hire 18-seater minibuses, how many minibuses do they need? What is the cost?

_____

(c) Which transport should they hire?

_____

3.4 Dividing 2-digit or 3-digit numbers by a 2-digit number (2)

Learning objective Use written methods to divide 3-digit numbers by 2-digit numbers

Basic questions

1 For each calculation, complete the statements then work out the answer. The first one has been done for you.

(a) $51\overline{)160}$

Think: When $16 \div 5$, the quotient is $\boxed{3}$.

We get: $\boxed{3} \times 51 = \boxed{153}$.

The quotient is ___just right___. (Choose: just right, too big or too small.)

(b) $63\overline{)480}$

Think: When $48 \div 6$, the quotient is $\boxed{}$.

We get: $\boxed{} \times 63 = \boxed{}$.

The quotient is _____. (Choose: just right, too big or too small.)

Change the quotient to $\boxed{}$.

$\boxed{} \times 63 = \boxed{}$.

The remainder is $\boxed{}$; it is _____ than the divisor.

Therefore, the quotient $\boxed{}$ is the right choice.

(c) $93\overline{)360}$

Think: When 36 ÷ 9, the quotient is ☐.

We get: ☐ × 93 = ☐.

The quotient is _____. (Choose: just right, too big or too small.)

Change the quotient to ☐.

☐ × 93 = ☐.

The remainder is ☐; it is _____ than the divisor.

Therefore, the quotient ☐ is the right choice.

(d) $43\overline{)334}$

Think: When 33 ÷ 4, the quotient is ☐.

We get: ☐ × 43 = ☐.

The quotient is _____. (Choose: just right, too big or too small.)

Change the quotient to ☐.

☐ × 43 = ☐.

The remainder is ☐; it is _____ than the divisor.

Therefore, the quotient ☐ is the right choice.

2 Use the column method to calculate.

505 ÷ 89 =

321 ÷ 56 =

317 ÷ 66 =

84 ÷ 22 =

178 ÷ 22 =

453 ÷ 57 =

3 What is the greatest number you can use to make each number statement correct?

30 × ⬚ < 207   60 × ⬚ < 415   40 × ⬚ < 316

70 × ⬚ < 566   50 × ⬚ < 410   80 × ⬚ < 768

4 Write the number sentences and then calculate.

(a) 192 is divided by 32. What is the quotient?

_____

(b) What times 43 is 344?

_____

(c) What is the quotient of 100 000 divided by 10?

_____

(d) What is the quotient of 100 000 divided by 100?

_____

5 A team of road maintenance workers are repairing an 855 metre-long road. The workers have completed 162 m of the road. For the remaining part, if they repair 63 m of the road each day, how many more days do they need to complete their work?

_____

_____

Challenge and extension question

6 A tunnel is 760 m long. A 240 m-long train is travelling at 25 m per second. How long will it take to pass through the tunnel?

_____

_____

3.5 Dividing 2-digit or 3-digit numbers by a 2-digit number (3)

 Learning objective Use written methods to divide 3-digit numbers by 2-digit numbers

 Basic questions

1 For each calculation, complete the statements then work out the answer.

(a) 28)‾89‾

Think: When 80 ÷ 20, the quotient is ⬚.

⬚ × 28 = ⬚ .

The quotient is _____ (Choose: just right, too big or too small); change the quotient to ⬚ .

⬚ × 28 = ⬚ .

The remainder is ⬚ ; it is _____ than the divisor.

Therefore, the quotient ⬚ is the right choice.

(b) 28)‾80‾

Think: When 80 ÷ 20, the quotient is ⬚.

⬚ × 28 = ⬚ .

The quotient is _____ (Choose: just right, too big or too small); subtract 1, it is ⬚ .

☐ × 28 = ☐ .

Again subtract 1, it is ☐ .

☐ × 28 = ☐ .

The remainder is ☐ ; it is _____ than the divisor.

Therefore, the quotient ☐ is the right choice.

(c) 38)‾2‾7‾8‾

Think: When 270 ÷ 30, the quotient is ☐ .

When 270 ÷ 40, the quotient is ☐ .

The quotient ☐ is the right choice.

(d) 57)‾4‾2‾1‾

Think: When 420 ÷ 50, the quotient is ☐ .

When 420 ÷ 60, the quotient is ☐ .

The quotient ☐ is the right choice.

(e) 19)‾1‾3‾4‾

Think: Both have the same first digit and 13 < 19.

Try the number ☐ as an initial quotient.

130 ÷ 20, the quotient is ☐ .

The quotient ☐ is the right choice.

(f) $78\overline{)732}$

Think: Both have the same first number and 73 < 78.

Try ⬚ as an initial quotient.

730 ÷ 80, the quotient is ⬚.

The quotient ⬚ is the right choice.

2 Choose the most appropriate method, mental or written, to find the answer to each calculation.

88 ÷ 22 =

72 ÷ 36 =

325 ÷ 38 =

188 ÷ 22 =

272 ÷ 36 =

829 ÷ 89 =

3 Write the number sentences and then calculate the answer.

(a) How many times 48 is 384?

_____

(b) 59 times a number is 413. What is the number?

_____

(c) At least how much needs to be taken away from 960 so there is no remainder when it is divided by 42?

_____

4 In a spring week, there were 1000 birds visiting an island in their migration from south to north. Among them, 130 were cuckoos. How many were other species of birds? Excluding the remainder, how many times as many as cuckoos were the other species of birds? What is the remainder?

_____

Challenge and extension question

5 Fill in the boxes with suitable numbers to complete each column division.

(a)

```
        □ □ 5
  63 ) □ □ □
      □ □ □
        □ 7
```

(b)

```
        □ 7
  1 □ ) □ □ 4
      □ □ □
      □ □ □ 2
```

(c)

```
            □
  2 □ ) 2 □ □
       □ □   8
       1   5
```

3.6 Dividing multi-digit numbers by a 2-digit number (1)

Learning objective Use written methods to divide multi-digit numbers by 2-digit numbers

Basic questions

1 Use the steps shown to help find the answer to each calculation.

(a) $204 \div 12$

$12 \times 10 = $ ☐

$12 \times 20 = $ ☐

First: ☐ $\div 12 = $ ☐

Then: $84 \div 12 = $ ☐

Therefore: $204 \div 12 = $ ☐

(b) $2128 \div 38$

$38 \times 50 = $ ☐

$38 \times 60 = $ ☐

First: ☐ $\div 38 = $ ☐

Then: ☐ $\div 38 = $ ☐

Therefore: $2128 \div 38 = $ ☐

2 Try the following divisions.

$20 \overline{)500}$

$30 \overline{)500}$

$40 \overline{)500}$

3 Use the column method to find the answer to each calculation.

999 ÷ 13 =

999 ÷ 23 =

999 ÷ 33 =

999 ÷ 43 =

999 ÷ 53 =

999 ÷ 83 =

4 Fill in the hundreds place in each dividend below with a number starting from 1 to 9 in order and then find the quotient and remainder. Identify patterns from these divisions.

(a)

35) ☐ 3 5   35) ☐ 3 5   35) ☐ 3 5

35) ☐ 3 5   35) ☐ 3 5   35) ☐ 3 5

35) ☐ 3 5   35) ☐ 3 5   35) ☐ 3 5

(b) Pattern identified:

When the hundreds place is filled with any of the

digits _____, the quotient is a 1-digit number.

When the hundreds place is filled in with either _____,
the quotient is a 2-digit number.

5 Fill in the missing numbers to make each statement correct.

(a) The quotient of 268 ÷ 26 is a ☐-digit number.

The highest value place of the quotient is in the _____ place.

The quotient is ☐.

(b) The quotient of 268 ÷ 38 is a ☐-digit number.

The digit of the quotient in the highest-value place is ☐.

(c) When the quotient of 5█6 ÷ 53 is a 2-digit number, the least possible

number in the █ is ☐.

(d) The highest-value place of the quotient of 336 ÷ 3█ is in the tens

place; the greatest possible number in the █ is ☐.

Challenge and extension question

6 Fill in the boxes with suitable numbers to complete the column division.

Left problem:

```
        □ 9
    ┌────────────
 26 )□ □ □
    2 □
    ───────────
      □ □ 6
      □ □ □
    ───────────
        2   2
```

Right problem:

```
          2 □
      ┌────────────
  □8 ) 9 □ 0
      □ □
    ───────────
      □ □ □
      □ □ □
    ───────────
            2
```

3.7 Dividing multi-digit numbers by a 2-digit number (2)

Learning objective Use mental and written methods to divide multi-digit numbers by 2-digit numbers

Basic questions

1 Do the following divisions.

$$60 \overline{\smash{)}2000}$$

$$24 \overline{\smash{)}2000}$$

$$12 \overline{\smash{)}2000}$$

$$16 \overline{\smash{)}2000}$$

2 Use the column method to calculate.

5000 ÷ 97 =

5000 ÷ 78 =

5000 ÷ 54 =

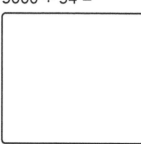

5000 ÷ 48 =

5000 ÷ 26 =

5000 ÷ 13 =

3 Work these out step by step. (Calculate smartly when possible.)

3265 + 806 − 265

35 × 9 + 265

18 × 25 + 25 × 22

3333 ÷ 11 × 4

625 ÷ 25 × 40

158 × 99 + 158

4 Write the number sentences and then calculate the answer.

(a) When the product of 84 and 14 is divided by 49, what is the quotient?

(b) How many times 18 is the sum of the greatest 3-digit number and the greatest 2-digit number?

(c) When the product of two fifteens is divided by 225, what is the quotient?

_____

_____

5 Multiple choice questions. (For each question, choose the correct answer and write the letter in the box.)

(a) In the division sentences below, the quotient that is a 2-digit number is ☐.

   A. 2188 ÷ 22   **B.** 324 ÷ 55   **C.** 843 ÷ 8   **D.** 3838 ÷ 28

(b) A 4-digit number is divided by a 2-digit number. The quotient is ☐.

   A. 2-digit number

   B. 3-digit number

   C. 1-digit number

   D. 2-digit number or 3-digit number

Challenge and extension question

6 Complete the column division calculations.

3.8 Practice and exercise

 Learning objective Solve division problems

 Basic questions

1 Work these out mentally and then write the answers.

(a) $35 \times 20 =$ ☐

(b) $36 \times 5 \div 18 =$ ☐

(c) $79 \times 0 \times 12 =$ ☐

(d) $6800 \div 34 =$ ☐

(e) $86 \times 5 - 30 =$ ☐

(f) $100 - 100 \div 25 =$ ☐

(g) $9300 \div 31 =$ ☐

(h) $3000 \div 25 =$ ☐

(i) $2000 \div 500 =$ ☐

(j) $1000 \div 250 =$ ☐

(k) $9000 \div 3000 =$ ☐

(l) $10\,000 \div 100 =$ ☐

2 Use the column method to calculate. Check the answers to the questions marked with *.

(a) $3025 \times 88 =$

(b) $*3296 \div 32 =$

(c) $*2551 \div 42 =$

  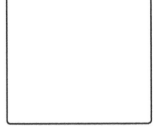

3 Work these out step by step. (Calculate smartly when possible.)

(a) 432 × 16 ÷ 12

(b) 32 × (275 − 75)

(c) 1033 × 22 + 189 × 9

(d) 2000 ÷ 25 ÷ 8

(e) (2216 + 888) ÷ 32

(f) 9676 ÷ 41 − 6524 ÷ 28

4 Write the number sentences and then calculate the answers.

(a) A is 4736 and it is 16 times B. What is the difference between A and B?

(b) The sum of 565 and 191 is divided by 18. What is the quotient?

5 Fill in the missing numbers to make each statement correct.

(a) Find the quotients of the following division calculations.

125 ÷ 25 = ☐   1250 ÷ 25 = ☐   1275 ÷ 25 = ☐

(b) When the quotient of 440 ÷ ■3 is a 2-digit number, the greatest possible number in the ■ is ☐ .

(c) When the quotient of 9▮95 ÷ 95 is a 3-digit number, the least

possible number in the ▮ is [ ] ; in this case, the

quotient is [ ] and the remainder is [ ] . When the quotient

is a 2-digit number, the greatest possible number in the ▮ is [ ] ;

in this case, the quotient is [ ] and the remainder is [ ] .

Challenge and extension question

6 The distance between City A and City B is 220 km. How long will it take
you to get there if you have the following means of transportation?

Means of transportation	Speed	Time
On foot	5 km/h	
By scooter	20 km/h	
By motorcycle	40 km/h	
By bus	80 km/h	
By train	110 km/h	
By high-speed train	275 km/h	

Chapter 3 test

1 Work these out mentally and then write the answers.

(a) $180 \div 15 = $

(b) $78 \times 10 = $

(c) $6300 \div 70 = $

(d) $1000 \div 25 = $

(e) $24 \times 50 = $

(f) $30 \times 99 = $

(g) $9300 \div 300 = $

(h) $10\,000 \div 200 = $

(i) $240 \div (7 + 3) = $

(j) $72 \div (86 - 74) = $

(k) $60 - 51 \div 3 = $

(l) $800 + 80 \div 40 = $

(m) $9600 \div 80 \div 30 = $

(n) $4 \times 6 \times 50 = $

(o) $75 + 25 \div 25 = $

(p) $(700 - 220) \div 60 = $

2 Use the column method to calculate. Check the answer to the question marked with *.

(a) $509 \times 92 = $

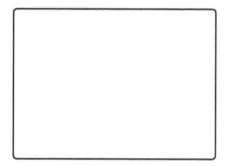

(b) $27 \times 6300 = $

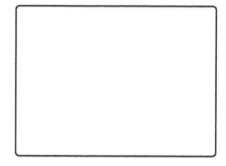

(c) $7380 \div 36 = $

(d) *$17\,052 \div 28 = $

3 Work these out step by step.

(a) 640 × 12 ÷ 80

(b) 9090 ÷ 15 × 20

(c) 1700 ÷ 25 ÷ 4

(d) 39 × 75 + 9898 ÷ 98

4 Complete each statement.

(a) Given 45 × 58 = 2610, 450 × 5800 = [ ].

(b) The quotient of 1190 divided by 17 is a [ ]-digit number.

The highest-value place of the quotient is in the _____ place.

(c) The quotient of 7548 ÷ 75 is a [ ]-digit number.

The highest-value place of the quotient is in the _____ place,

there are [ ] zeros in the end and the remainder is [ ].

(d) To make the quotient of 784 ÷ 47 a 3-digit number, the digit in

the  could be [ ].

(e) In 32)‾238, if the quotient is a 3-digit number, the least possible

number in the  is [ ]. If the quotient is a 2-digit number, the

number in the  could be [ ].

5 Multiple choice questions. (For each question, choose the correct answer and write the letter in the box.)

(a) Bob, Hannah, Kate and Mark were having a 60-metre run. Bob took 14 seconds, Hannah took 13 seconds, Kate took 15 seconds and Mark took 16 seconds. The fastest runner was ☐.

    A. Bob   **B.** Hannah   **C.** Kate   **D.** Mark

(b) The number sentence that has a different answer from the calculation of 84×46 is ☐.

    A. $(80 + 4) \times 46$ **B.** $84 \times 40 + 84 \times 6$

    C. $80 \times 46 + 4$ **D.** $80 \times 46 + 4 \times 46$

(c) A 4-digit number is divided by a 2-digit number; the quotient is a ☐-digit number.

    A. 4   **B.** 3   **C.** 3 or 2   **D.** Not sure

(d) In $54\,081 \div 27$, there are ☐ zeros in the middle of the quotient.

    A. 0   **B.** 1   **C.** 2   **D.** 3

(e) In $605\,605 \div 605$, there are ☐ zeros in the middle of the quotient.

    A. 1   **B.** 2   **C.** 3   **D.** 4

6 Solve these problems.

(a) Evan read 84 pages of a book in 3 days. The book has 504 pages. If he reads at the same speed, in how many days can he read the whole book?

_____

_____

(b) A clothing factory made 60 sets of children's clothes each day on average in the first 3 days of the week. In the next 4 days, it made 310 sets. How many sets of children's clothes has it made altogether in that week?

(c) James walks 5 km per hour. If he cycles he can be 10 km faster per hour than he walks. How many kilometres can James travel if he cycles for 3 hours?

(d) There are 4500 bottles of juice in a supermarket. 125 boxes of juice were sold. Each box has 24 bottles. How many bottles of juice were left?

7 Given that the distance from Shanghai to another Chinese city is 1328 km, find the speeds of the following means of transportation. Fill in the table with your answers (rounding to the nearest whole number).

Means of transportation	Speed	Time
By regular train	_____ km/h	9 hours and 29 minutes
By high-speed train	_____ km/h	5 hours and 18 minutes
By aeroplane	_____ km/h	1 hour and 39 minutes

Chapter 4 Comparing fractions, improper fractions and mixed numbers

4.1 Comparing fractions (1)

Learning objective Compare and order fractions with the same denominator

Basic questions

1 Use the diagram to complete the statements.

(a)

$\frac{1}{7}$	$\frac{1}{7}$	$\frac{1}{7}$	$\frac{1}{7}$	$\frac{1}{7}$	$\frac{1}{7}$	$\frac{1}{7}$

$\frac{1}{7}$	$\frac{1}{7}$	$\frac{1}{7}$	$\frac{1}{7}$	$\frac{1}{7}$	$\frac{1}{7}$	$\frac{1}{7}$

$\frac{3}{7}$ means ⬚ lots of $\frac{1}{7}$. $\frac{6}{7}$ means ⬚ lots of $\frac{1}{7}$.

Therefore $\frac{3}{7}$ is _____ than $\frac{6}{7}$. (Choose 'greater' or 'less'.)

(b) For fractions that have the same denominator, the greater the

numerator is, the _____ the fraction is, and the smaller the

_____ is, the smaller the fraction is.

2 Use fractions to represent the shaded parts of the figures and fill in the

◯ with >, < or =.

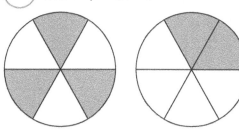

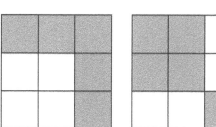

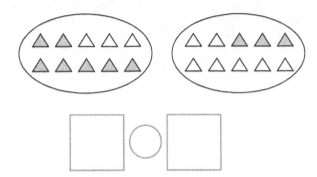

3 Colour $\frac{3}{8}$, $\frac{7}{8}$ and $\frac{1}{8}$ in the following diagram. Write them in order below, starting from the smallest.

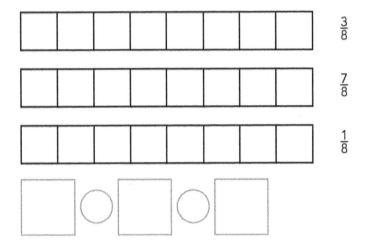

$\frac{3}{8}$

$\frac{7}{8}$

$\frac{1}{8}$

4 Compare the fractions and write >, < or = in the ◯.

(a) $\frac{3}{10}$ ◯ $\frac{8}{10}$

(b) $\frac{6}{7}$ ◯ $\frac{1}{7}$

(c) $\frac{6}{9}$ ◯ $\frac{9}{9}$

(d) $\frac{5}{5}$ ◯ $\frac{4}{5}$

(e) $\frac{9}{20}$ ◯ $\frac{18}{20}$

(f) $\frac{53}{109}$ ◯ $\frac{77}{109}$

(g) $\frac{9}{10}$ ◯ $\frac{7}{10}$ ◯ $\frac{1}{10}$

(h) $\frac{1}{8}$ ◯ $\frac{5}{8}$ ◯ $\frac{8}{8}$

(i) $\frac{13}{18}$ ◯ $\frac{7}{18}$ ◯ $\frac{9}{18}$

Challenge and extension questions

5 Put the fractions $\frac{7}{9}$, $\frac{2}{9}$, $\frac{1}{9}$, $\frac{5}{9}$ and $\frac{3}{9}$ in order, from the smallest to the greatest.

6 Put the fractions $\frac{30}{80}$, $\frac{1}{80}$, $\frac{18}{80}$, $\frac{79}{80}$ and $\frac{50}{80}$ in order, from the greatest to the smallest.

7 Four groups of children in a Year 5 class borrowed 39 skipping ropes from the school sports shed. The first group received 8, the second group received 10, the third group got 11, and the fourth group received all of the remaining ones.

What fraction of the ropes did the fourth group borrow? _____

4.2 Comparing fractions (2)

Learning objective Compare and order fractions with a numerator of 1

Basic questions

1 Use the diagram to complete the statements.

| $\frac{1}{6}$ | $\frac{1}{6}$ | $\frac{1}{6}$ | $\frac{1}{6}$ | $\frac{1}{6}$ | $\frac{1}{6}$ |

| $\frac{1}{3}$ | $\frac{1}{3}$ | $\frac{1}{3}$ |

| $\frac{1}{2}$ | $\frac{1}{2}$ |

(a) $\frac{1}{6}$ is _____ than $\frac{1}{3}$. (Choose 'greater' or 'less'.)

(b) $\frac{1}{3}$ is _____ than $\frac{1}{2}$. (Choose 'greater' or 'less'.)

(c) For fractions that have the same numerator, the greater the

denominator is, the _____ the fraction is; the smaller the

_____ is, the greater the fraction is.

2 Fill in the spaces.

(a) Use a fraction to represent the shaded part in each figure below.

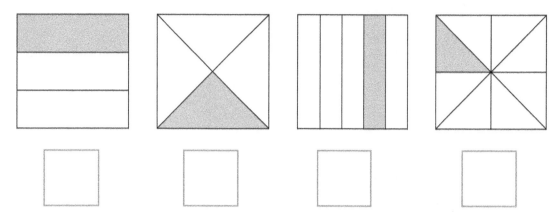

(b) Put the fractions from part (a) in order, starting from the greatest.

(c) From the above, we found that if the whole is the same, the more

parts the whole is equally divided into, the _____ each
part gets. Therefore, for a fraction with a numerator of 1, the greater

the denominator is, the _____ the fraction is.

3 Compare the fractions and write >, < or = in the $\bigcirc$.

(a) $\frac{1}{10} \bigcirc \frac{1}{8}$

(b) $\frac{1}{6} \bigcirc \frac{1}{7}$

(c) $\frac{1}{9} \bigcirc \frac{1}{15}$

(d) $\frac{1}{4} \bigcirc \frac{1}{40}$

(e) $\frac{1}{21} \bigcirc \frac{1}{20}$

(f) $\frac{1}{109} \bigcirc \frac{1}{77}$

(g) $\frac{1}{9} \bigcirc \frac{1}{7} \bigcirc \frac{1}{5}$

(h) $\frac{1}{8} \bigcirc \frac{1}{18} \bigcirc \frac{1}{108}$

Challenge and extension questions

4 Put the fractions $\frac{1}{9}$, $\frac{1}{36}$, $\frac{1}{27}$, $\frac{1}{18}$ and $\frac{1}{81}$ in order, from the smallest to the greatest.

☐ ☐ ☐ ☐ ☐

5 Put the fractions $\frac{1}{10}$, $\frac{1}{20}$, $\frac{1}{70}$, $\frac{1}{100}$ and $\frac{1}{40}$ in order, from the greatest to the smallest.

☐ ☐ ☐ ☐ ☐

6 There are two boxes of chocolates. The first box has 30 chocolates and the second has 20 chocolates. Alex takes $\frac{1}{5}$ of the chocolates from the first box and his brother takes $\frac{1}{4}$ of the chocolates from the second box.

Who takes more? _____

4.3 Comparing fractions (3)

 Learning objective Compare and order fractions with the same numerator and different denominators

 Basic questions

1 (a) Write a fraction to represent the shaded part in each figure.

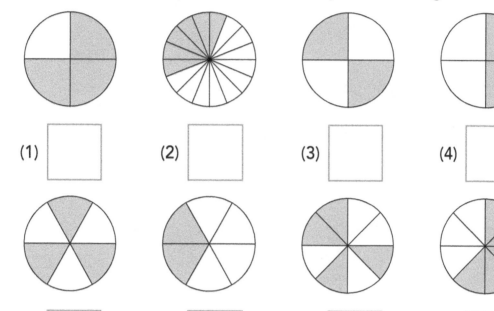

(1) ☐   (2) ☐   (3) ☐   (4) ☐

(5) ☐   (6) ☐   (7) ☐   (8) ☐

(b) Compare: The figures that have the same portion of shaded parts as figure (3) are _____.

(c) Comparing figure (1), figure (2) and figure (6) we can see that if the whole is the same, the fewer parts the whole is equally divided into,

the _____ each part is, and for fractions with the same

numerator, the _____ the denominator is, the greater the fraction is.

2 Compare the fractions and write >, < or = in the ◯.

(a) $\frac{1}{63}$ ◯ $\frac{1}{36}$

(b) $\frac{3}{20}$ ◯ $\frac{3}{80}$

(c) $\frac{8}{100}$ ◯ $\frac{8}{99}$

(d) $\frac{7}{41}$ ◯ $\frac{7}{40}$

(e) $\frac{25}{200}$ ◯ $\frac{25}{300}$

(f) $\frac{11}{535}$ ◯ $\frac{11}{553}$

(g) $\frac{60}{601}$ ◯ $\frac{60}{699}$ ◯ $\frac{60}{700}$

3 Fill in the missing numbers to make each statement correct.

(a) There are ☐ lots of $\frac{1}{15}$ in $\frac{7}{15}$. After taking away 3 lots of $\frac{1}{15}$, it is ☐.

(b) Five lots of $\frac{1}{13}$ make ☐. Four lots of $\frac{1}{13}$ make ☐.

The difference between them is ☐.

(c) A 2 metre-long piece of string is cut into 5 equal pieces. Each piece is ☐ of the string and it is ☐ m long.

Four pieces are ☐ of the string, and the length of each piece is ☐ m.

(d) There are 19 boys in a maths class, which is $\frac{19}{37}$ of the class.

The number of girls in the class is ☐.

Challenge and extension questions

4 Put the fractions $\frac{5}{123}$, $\frac{5}{999}$, $\frac{5}{656}$, $\frac{5}{50}$ and $\frac{5}{11}$ in order, from the smallest to the greatest.

5 Put the fractions $\frac{4}{7}$, $\frac{2}{11}$ and $\frac{2}{7}$ in order, from the greatest to the smallest.

6 A snail and an ant had a race to climb a wall. The snail climbed $\frac{9}{15}$ m and the ant climbed $\frac{9}{20}$ m. Who climbed higher? _____

7 Tom and Mary had two cups of drinks with the same amount in each. Tom drank $\frac{3}{4}$ of his and Mary drank $\frac{2}{3}$ of hers.

Who drank more? _____

4.4 Comparing fractions (4)

Learning objective Compare equivalent fractions

Basic questions

1 Read the story and then answer the questions.

> Once upon a time, an old monkey gave some young monkeys some peaches to share. The old monkey gave one of the young monkeys a basket of peaches and said, 'These are for six of you to share.'
>
> The young monkey thought it was too few and asked for more. The old monkey said, 'OK, you can have one more basket, but to share with 12 monkeys.' The young monkey was very happy with this.

(a) Given that each basket contains the same number of peaches, what do you think about the story?

_____

_____

_____

_____

Lee thinks:

Six monkeys share a basket of peaches. Each monkey can have $\frac{1}{6}$ of the peaches. Two baskets of peaches are shared by 12 monkeys. Each monkey has $\frac{2}{12}$ of the peaches.

Ming thinks:

As learned earlier, we know different fractions can represent the same part of an object, and $\frac{1}{6}$ and $\frac{2}{12}$ are equal. Therefore, the number of peaches that each monkey has is the same either way.

(b) Do you think Lee or Ming is right? Why?

_____

_____

_____

_____

2 Write equivalent fractions.

(a) $\frac{1}{2} = \frac{2}{\boxed{}} = \frac{4}{\boxed{}}$

(b) $\frac{1}{5} = \frac{2}{\boxed{}} = \frac{\boxed{}}{40}$

(c) $\frac{6}{15} = \frac{\boxed{}}{5} = \frac{4}{\boxed{}}$

(d) $\frac{5}{5} = \frac{4}{\boxed{}} = \frac{\boxed{}}{99}$

(e) $\frac{6}{18} = \frac{12}{\boxed{}} = \frac{\boxed{}}{9} = \frac{24}{\boxed{}} = \frac{1}{\boxed{}} = \frac{2}{\boxed{}}$

(f) $1 = \frac{11}{11} = \frac{33}{\boxed{}} = \frac{\boxed{}}{300}$

3 Complete each statement.

(a) Mary was given half of half a cake; she was given ☐ of the cake.

(b) Write any three fractions that are equal to $\frac{1}{3}$: ☐ , ☐

and ☐ .

(c) 3 lots of $\frac{1}{6}$ equal ☐ lots of $\frac{1}{12}$

(d) After folding a square piece of paper in half twice, each part is ☐ of the square.

4 An ant was crawling from Place A to Place B.

The ant crawled $\frac{4}{12}$m on the first day and $\frac{6}{12}$m on the second day.

$\frac{4}{12}$m $\frac{6}{12}$m

A   B

1 m

(a) How many metres did the ant crawl on these two days?

_____

(b) How many metres away is it to Place B after two days' crawling?

_____

Challenge and extension questions

5 Walking from Place A to Place B took Tim $\frac{3}{7}$ hours and Sam $\frac{4}{14}$ hours.

Who walked faster? _____

6 Three pizzas of the same size are shared equally by four children.

How many pizzas can each child have? Show your working. _____

4.5 Improper fractions and mixed numbers

Learning objective Convert between mixed numbers and improper fractions

Basic questions

1 Use the following numbers and fractions to represent the shaded part in each figure.

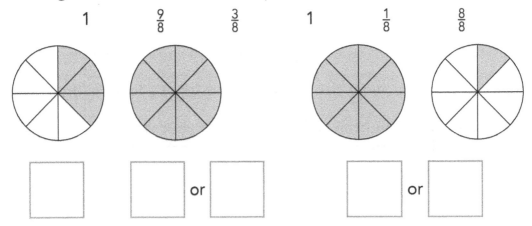

$$1 \qquad \frac{9}{8} \qquad \frac{3}{8} \qquad 1 \qquad \frac{1}{8} \qquad \frac{8}{8}$$

| | | or | | | or | |

2 Sort these fractions into proper fractions, improper fractions and mixed numbers.

$$\frac{7}{12} \quad \frac{5}{3} \quad 7\frac{1}{18} \quad \frac{3}{2} \quad 1\frac{4}{5} \quad \frac{79}{100} \quad 30\frac{1}{2} \quad \frac{181}{365} \quad \frac{13}{12} \quad \frac{19}{6} \quad 5\frac{1}{4}$$

Proper fractions: _____

Improper fractions: _____

Mixed numbers: _____

3 Mark the following numbers on the number line and then write them in order, from the smallest to the greatest.

0   10

$a = 2\frac{1}{2}$ $b = \frac{3}{4}$ $c = 8\frac{2}{3}$ $d = 5$ $e = 9\frac{1}{3}$

From the smallest to the greatest: _____

_____

4 Read these statements and write 'true' or 'false'.

(a) $2\frac{3}{5}$ is read as two and three fifths.   _____

(b) All proper fractions are less than 1.   _____

(c) All improper fractions are greater than 1.   _____

(d) The numerator of a proper fraction is always less than

its denominator.   _____

(e) The numerator of an improper fraction is always greater

than its denominator.   _____

(f) A proper fraction can be converted to a mixed number.   _____

(g) An improper fraction can be converted to a mixed number. _____

(h) A mixed number can be converted to an improper fraction. _____

5 Convert the mixed numbers to improper fractions.

(a) $2\frac{1}{3} = \boxed{}$ (b) $1\frac{5}{8} = \boxed{}$ (c) $6\frac{7}{9} = \boxed{}$ (d) $60\frac{37}{39} = \boxed{}$

6 Convert the improper fractions to whole numbers or mixed numbers.

(a) $\frac{31}{7} = $ _____

(b) $\frac{53}{8} = $ _____

(c) $\frac{81}{9} = $ _____

(d) $\frac{97}{30} = $ _____

 Challenge and extension questions

7 Complete the following table. One has been done for you.

Quantity	As a decimal	As a mixed number	As an improper fraction
2 m 10 cm	2.1 m	$2\frac{1}{10}$ m	$\frac{21}{10}$ m
90 minutes	_____ h	_____ h	_____ h
5 kg 200 g	_____ kg	_____ kg	_____ kg
1650 ml	_____ l	_____ l	_____ l
30 km 200 m	_____ km	_____ km	_____ km

8 (a) If $\frac{5}{\triangle + 2}$ is an improper fraction and $\triangle$ is a whole number (including 0), then $\triangle$ can be _____.

(b) If a is a whole number, $\frac{a}{9}$ is a proper fraction, and $\frac{a}{5}$ is an improper fraction, then the value of a can be _____.

4.6 Adding and subtracting fractions with related denominators (1)

Learning objective Add and subtract fractions with related denominators

Basic questions

1 Draw a line to link each pair of equivalent fractions.

$\frac{1}{2}$ $\qquad$ $\frac{2}{5}$ $\qquad$ $\frac{13}{18}$ $\qquad$ $\frac{11}{10}$ $\qquad$ $\frac{17}{20}$

$\frac{65}{90}$ $\qquad$ $\frac{85}{100}$ $\qquad$ $\frac{3}{6}$ $\qquad$ $\frac{4}{10}$ $\qquad$ $\frac{110}{100}$

2 Fill in the missing numbers.

(a) $\frac{1}{3} = \frac{\boxed{}}{6}$

(b) $\frac{2}{5} = \frac{\boxed{}}{10}$

(c) $\frac{17}{10} = \frac{34}{\boxed{}}$

(d) $\frac{13}{100} = \frac{\boxed{}}{200}$

(e) $\frac{31}{25} = \frac{93}{\boxed{}}$

(f) $\frac{10}{17} = \boxed{}$

3 Work these out mentally. Write the answers.

(a) $\frac{1}{5} + \frac{1}{5} = \boxed{}$

(b) $\frac{4}{7} - \frac{2}{7} = \boxed{}$

(c) $\frac{3}{10} + \frac{7}{10} = \boxed{}$

(d) $\frac{9}{13} - \frac{5}{13} = \boxed{}$

(e) $\frac{8}{9} - \frac{4}{9} = \boxed{}$

(f) $\frac{13}{23} + \frac{5}{23} = \boxed{}$

Comparing fractions, improper fractions and mixed numbers

4 Add the fractions step by step and write your answers in mixed numbers if they are greater than 1. The first one has been done for you.

(a) $\frac{1}{3} + \frac{5}{6}$

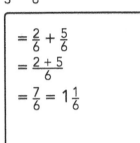
$$= \frac{2}{6} + \frac{5}{6}$$
$$= \frac{2+5}{6}$$
$$= \frac{7}{6} = 1\frac{1}{6}$$

(b) $\frac{3}{5} + \frac{1}{10} =$

(c) $\frac{9}{13} + \frac{7}{26} =$

(d) $\frac{1}{3} + \frac{13}{15} =$

(e) $\frac{5}{7} + \frac{18}{35} =$

(f) $\frac{20}{11} + \frac{51}{77} =$

5 Subtract the fractions step by step. The first one has been done for you.

(a) $\frac{1}{2} - \frac{3}{8}$

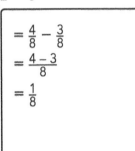
$$= \frac{4}{8} - \frac{3}{8}$$
$$= \frac{4-3}{8}$$
$$= \frac{1}{8}$$

(b) $\frac{2}{3} - \frac{5}{9} =$

(c) $\frac{15}{22} - \frac{4}{11} =$

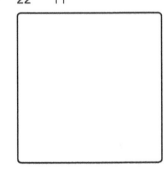

(d) $\frac{5}{7} - \frac{5}{14} =$

(e) $\frac{61}{75} - \frac{113}{150} =$

(f) $\frac{59}{90} - \frac{4}{9} =$

6 When Andrew said how many books he had, Jason said that he had two thirds as many as Andrew, and his younger sister Amy said she had one ninth as many as Andrew. Answer the following questions. Show your working.

(a) Comparing the numbers of books Jason and Amy have, who has more books? _____

(b) Express the number of books Jason and Amy have in total as a fraction of what Andrew has. ☐

(c) Express the difference between the numbers of books Jason and Amy have as a fraction of what Andrew has. ☐

(d) If Andrew has 18 books, how many books do Jason and Amy have altogether? (Use two different ways to find the answer.)

Method 1:

Method 2:

Challenge and extension questions

7 Calculate the following.

(a) $\frac{1}{2} + \frac{1}{4} + \frac{1}{8} =$

(b) $1 - \frac{1}{3} - \frac{1}{6} =$

8 Which of the following is correct? _____

A. $\frac{1}{2} + \frac{1}{3} = \frac{1}{5}$ **B.** $\frac{1}{2} + \frac{1}{3} = \frac{2}{5}$ **C.** $\frac{1}{2} + \frac{1}{3} = \frac{1}{6}$ **D.** $\frac{1}{2} + \frac{1}{3} = \frac{5}{6}$

Explain how you got the answer.

_____

_____

4.7 Adding and subtracting fractions with related denominators (2)

 Learning objective Add and subtract fractions, including mixed numbers, with related denominators

 Basic questions

1 Work these out mentally. Write the answers.

(a) $\frac{1}{7} + \frac{4}{7} =$ ☐

(b) $\frac{1}{5} - \frac{1}{5} =$ ☐

(c) $\frac{3}{2} + \frac{1}{8} =$ ☐

(d) $\frac{8}{9} - \frac{1}{3} =$ ☐

(e) $\frac{5}{7} + \frac{7}{56} =$ ☐

(f) $1 - \frac{9}{10} =$ ☐

2 Write the answers as a mixed number.

(a) $1 + \frac{1}{2} =$ ☐

(b) $15 + \frac{3}{7} =$ ☐

(c) $\frac{23}{100} + 58 =$ ☐

3 Find the difference and write the answer as a mixed number if it is greater than 1.

(a) $1 - \frac{1}{3} =$ ☐

(b) $1 - \frac{2}{15} =$ ☐

(c) $5 - \frac{17}{100} =$ ☐

4 Write the answer as a mixed number if it is greater than 1.

(a) $\frac{10}{7} + \frac{22}{7} =$ ☐

(b) $\frac{5}{6} + \frac{4}{3} =$ ☐

(c) $\frac{19}{7} - \frac{15}{7} =$ ☐

(d) $\frac{145}{99} - \frac{40}{33} =$ ☐

(e) $1 - \frac{2}{5} + \frac{7}{25} =$ ☐

(f) $\frac{4}{9} + \frac{23}{18} =$ ☐

5 Complete the following addition and subtraction of fractions. Write the answer as a mixed number if it is greater than 1. One has been done for you.

(a) $2\frac{3}{7} + 1\frac{5}{7}$

$$= (2 + \tfrac{3}{7}) + (1 + \tfrac{5}{7})$$
$$= (2 + 1) + (\tfrac{3}{7} + \tfrac{5}{7})$$
$$= 3 + \tfrac{8}{7}$$
$$= 3 + 1\tfrac{1}{7}$$
$$= 4\tfrac{1}{7}$$

(b) $10\frac{10}{13} - 5\frac{3}{13}$

(c) $3\frac{4}{9} - 3\frac{5}{18}$

(d) $12\frac{2}{3} - 10\frac{7}{12}$

6 A number has $\frac{5}{6}$ subtracted from it and then has $\frac{1}{3}$ added to it; the answer is $\frac{2}{3}$. What is the start number? _____

7 An electrician used some electrical wire to wire a new building. On the first day he planned to use $1\frac{2}{5}$m of the wire, on the second day he planned to use $\frac{3}{5}$m more than on the first day, and on the third day he planned to use 2m more than on the second day. If the wire was 10m long, would it have been long enough? Why or why not?

_____

_____

 Challenge and extension question

8 While working on addition and subtraction involving mixed numbers, Dan found another method to find the answer, that is, first convert all the mixed numbers to improper fractions and then add or subtract them. Can you use Dan's method to find the answers to the following questions? Show your working.

(a) $9\frac{6}{19} + \frac{11}{19} =$

(b) $3\frac{2}{3} - 2\frac{7}{9} =$

4.8 Multiplying fractions by whole numbers

Learning objective Multiply fractions and mixed numbers by whole numbers

Basic questions

1 Fill in the missing numbers.

(a) $2 + 2 + 2 + 2 + 2 = \boxed{} \times \boxed{} = \boxed{}$

(b) $\frac{2}{7} + \frac{2}{7} = \dfrac{\boxed{} + \boxed{}}{7} = \dfrac{\boxed{}}{7}$

(c) $\frac{3}{10} + \frac{3}{10} + \frac{3}{10} = \dfrac{\boxed{} + \boxed{} + \boxed{}}{10} = \dfrac{\boxed{}}{10}$

(d) $\frac{1}{3} + \frac{1}{3} + \frac{1}{3} = \dfrac{\boxed{} + \boxed{} + \boxed{}}{3} = \boxed{}$

(e) $\frac{5}{9} + \frac{5}{9} = \dfrac{\boxed{} + \boxed{}}{9} = \dfrac{\boxed{}}{9} = \boxed{}$

(f) $\frac{4}{5} + \frac{4}{5} + \frac{4}{5} = \dfrac{\boxed{} + \boxed{} + \boxed{}}{5} = \boxed{}$

2 In each question below, one circle represents one whole. Use multiplication as repeated addition to find the total amount of shaded parts. Write your answer as a mixed number if it is greater than 1. The first one has been done for you.

(a)

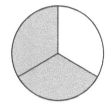

Total shaded parts: $4 \times \frac{2}{3} = \frac{2}{3} + \frac{2}{3} + \frac{2}{3} + \frac{2}{3} = \frac{2+2+2+2}{3} = \frac{4 \times 2}{3} = \frac{8}{3} = 2\frac{2}{3}$

(b)

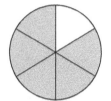 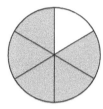 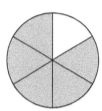 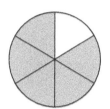

Total shaded parts: _____

(c)

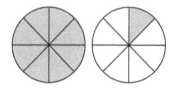

Total shaded parts: _____

3 Read the statements and write 'true' or 'false'.

(a) $5 \times \frac{2}{7} = \frac{5 \times 2}{5 \times 7} = \frac{10}{35}$ _____

(b) $3 \times 6\frac{3}{5} = (3 \times 6)\frac{3}{5} = 18\frac{3}{5}$ _____

(c) $3 \times \frac{7}{10} = \frac{3 \times 7}{10} = \frac{21}{10} = 2\frac{1}{10}$ _____

(d) $4 \times 2\frac{2}{9} = 4 \times (2 + \frac{2}{9}) = 4 \times 2 + 4 \times \frac{2}{9} = 8 + \frac{8}{9} = 8\frac{8}{9}$ _____

4 Multiplying fractions by whole numbers.

(a) $5 \times \frac{1}{2} =$

(b) $\frac{3}{10} \times 3 =$

(c) $8 \times 5\frac{2}{15} + \frac{8}{15} =$

(d) $12 \times 12\frac{1}{5} =$

(e) $3\frac{5}{12} \times 5 - 4 \times 2\frac{1}{12} =$

(f) $0 \times 6\frac{11}{100} + 1 \times 4\frac{19}{100} - 2 \times 2\frac{3}{100} =$

5 A rectangular pool is 9 metres long and $5\frac{1}{2}$ metres wide. Find its perimeter and area.

6 On a trip, Joshua drove from Cardiff to London and then to Edinburgh. It took him two and three-quarter hours to drive from Cardiff to London on the first day. On the second day, he drove from London to Edinburgh. The journey took him three times longer than the journey from Cardiff to London.

(a) How long did it take Joshua to drive from London to Edinburgh?

(b) How much longer did it take Joshua to drive from London to Edinburgh than from Cardiff to London?

(c) How much time did it take Joshua to drive from Cardiff to London and then from London to Edinburgh?

Challenge and extension question

7 Calculate the following and write the answer as a mixed number if it is greater than 1.

(a) $7 \times 5\frac{1}{2} + \frac{5}{6} =$

(b) $10 \times 3\frac{11}{12} - 12 \times 2\frac{3}{60} =$

Chapter 4 test

1 Work these out mentally. Write the answers.

(a) $12 \times 11 - 12 = \boxed{}$

(b) $100 \times 9 \div 30 = \boxed{}$

(c) $34 \div 17 + 305 = \boxed{}$

(d) $\frac{2}{5} + \frac{1}{5} = \boxed{}$

(e) $\frac{9}{15} - \frac{2}{15} = \boxed{}$

(f) $\frac{5}{7} + \frac{3}{14} = \boxed{}$

(g) $5 + \frac{3}{8} = \boxed{}$

(h) $4 \times \frac{4}{17} = \boxed{}$

(i) $10 \times 2\frac{1}{10} = \boxed{}$

2 Work these out step by step. Write the answer as a mixed number when it is greater than 1.

(a) $\frac{2}{3} + \frac{7}{15} =$

(b) $\frac{8}{13} - \frac{20}{39} =$

(c) $2 + \frac{3}{4} + \frac{25}{48} =$

(d) $3 + 2 \times \frac{5}{6} =$

(e) $4 \times \frac{7}{11} - 3 \times \frac{15}{22} =$

(f) $5 \times \frac{2}{3} - 2 \times \frac{5}{6} + 4 \times \frac{7}{18} =$

(g) $11 + 4 \times 3\frac{4}{11} =$

(h) $7 \times 5\frac{4}{9} - 11 \times 2\frac{17}{18} =$

3 Complete the statements and compare the fractions by writing >, < or = in the ◯.

(a) Colour the portion of each circle to represent the fractions given and then compare the fractions.

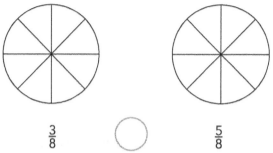

$\frac{3}{8}$ ◯ $\frac{5}{8}$

(b) Write the fractions of the shaded parts shown in the squares and then compare the fractions.

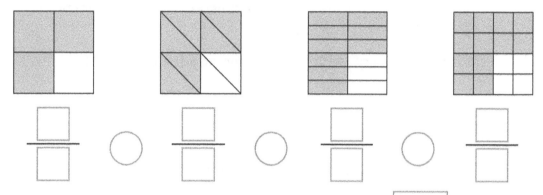

(c) The number representing three and one quarter is ☐ ;

doubling it is ☐ .

(d) The number representing three one-quarters is ☐ × ☐

or ☐ ; doubling it is ☐ .

4 Compare the fractions using the symbols >, < or = .

(a) $\frac{1}{2}$ ◯ $\frac{1}{3}$ (b) $\frac{5}{11}$ ◯ $\frac{6}{11}$ (c) $\frac{4}{102}$ ◯ $\frac{4}{105}$

(d) $\frac{3}{12}$ ◯ $\frac{1}{4}$ (e) $\frac{10}{30}$ ◯ $\frac{12}{30}$ (f) $3\frac{1}{10}$ ◯ $2\frac{9}{10}$

5 Put the fractions in order, starting from the greatest.

(a) $\frac{8}{12}, \frac{4}{12}, \frac{4}{14}, \frac{1}{14}$

(b) $\frac{1}{2}, 1, \frac{2}{8}, \frac{3}{4}$

6 Multiple choice questions. (For each question, choose the correct answer and write the letter in the box.)

(a) A teacher assigned 10 mental calculation problems. Within the same amount of time, Group A finished $\frac{7}{10}$ of the questions and Group B finished $\frac{1}{2}$ of them.

Which group solved the problems at a faster pace?

   A. Group A   **B.** Group B

   C. They have the same pace   **D.** Cannot be decided

(b) The incorrect statement of the following is ____.

   A. A mixed number is always greater than 1.

   B. The sum of any two mixed numbers is always greater than 1.

   C. A mixed number minus a proper fraction is always greater than 1.

   D. Multiplying a mixed number by a whole number (except zero) is always greater than 1.

7 Solve these problems.

(a) Ben and Jay walked from the local library to the school at the same time. It took Ben $\frac{2}{5}$ hours, while it took Jay $\frac{2}{3}$ hours.

Who walked faster? _____

(b) Stephanie's mother bought $2\frac{1}{2}$kg of oranges, 3kg of apples and $1\frac{1}{2}$kg of cherries. What is the total mass of the fruit she

bought? _____

(c) A willow tree was $\frac{4}{5}$m tall when it was planted. The height of the tree grew $1\frac{3}{10}$m in the first year. In the second year, it grew twice as much as in the first year.

(i) How much taller did the willow tree grow in the first two years?

(ii) What was the height of the tree at the end of the second year?

(d) The width of a rectangular lawn is $5\frac{9}{10}$m. Its length is 3 times its width.

(i) What is the length of the rectangular lawn?

(ii) How much longer is the length than the width?

(iii) What is the perimeter?

Chapter 5 Consolidation and enhancement

5.1 Large numbers and rounding (1)

 Learning objective Read, write and round large numbers

 Basic questions

1 Fill in the spaces to make each statement correct.

(a) Counting in 100 000s.

Forwards: 11 000, 111 000, 211 000, 311 000, ⬚ , ⬚ , 611 000

Backwards: 1 000 000, ⬚ , 800 000, ⬚ , ⬚ , 500 000

(b) When counting large numbers, ⬚ ten thousands is one hundred thousand; ⬚ ten millions is one hundred million.

(c) Ones, tens, hundreds, thousands, ten thousands, hundred thousands, millions, ten millions, hundred millions, billions, ten billions, are all units

of _____. The multiplication from one unit to the next unit is by 10.

(d) The two nearest whole million numbers to 1 567 000 are

⬚ and ⬚ .

(e) 43 007 070 is read in words as _____

_____.

It consists of [   ] thousands and [   ] ones.

(f) In a 3-digit number, the digit in the tens place is 1 greater than the digit in the ones place and 3 greater than the digit in the hundreds place. The digit in the hundreds place is half of 10.

The number is [   ].

2 Read out and write the following numbers in words.

(a) 4 204 322 _____

_____

(b) 10 025 090 _____

_____

(c) 20 000 200 _____

_____

(d) 1 010 101 010 _____

_____

3 Write the following numbers in numerals.

(a) Fifty-nine million, five hundred and eight thousand, eight hundred and eighty. [   ]

(b) Four hundred million, eight hundred and fifty-four thousand, five hundred. [   ]

(c) Two million, six hundred and sixteen thousand, three hundred and twenty-nine. [   ]

(d) One hundred and ten million, four hundred and nine thousand and eleven. ⬚

4 Multiple choice questions. (For each question, choose the correct answer and write the letter in the box.)

(a) The number consisting of 3 ten-millions, 6 millions and 9 thousands is ⬚.

A. 369 000　　B. 3 609 000　　C. 36 009 000　　D. 30 609 000

(b) In a large number, from the ones place to the left, the fourth place is ⬚; the place to the right of the hundred millions place is ⬚.

A. the thousands place　　　　　　B. the ten thousands place

C. the hundred thousands place　　D. the ten millions place

(c) ⬚ different 4-digit numbers can be formed using the digits 0, 1, 5 and 9.

A. 16　　　　　B. 18　　　　　C. 20　　　　　D. 24

5 Compare each pair of numbers below. Write > or < in the ◯.

(a) 9676 ◯ 9767

(b) 800 100 ◯ 810 000

(c) 4 406 000 ◯ 446 000

(d) 559 999 ◯ 590 000

(e) 900 009 ◯ 900 090

(f) 100 001 ◯ 10 001

6 Round the numbers and complete the table.

	To the nearest ten thousand	To the nearest hundred thousand	To the nearest ten million
36 890 700			
98 970 076			
109 827 000			

Challenge and extension questions

7 Use 8, 3, 5 and three 0s to form 6-digit numbers according to the conditions given.

(a) The greatest number is [　　　　]; rounding it to the nearest

ten thousand, it is [　　　　].

(b) The least number is [　　　　]; rounding it to the nearest

hundred thousand, it is [　　　　].

(c) The numbers with three zeros at the end are _____

_____

_____.

(d) The numbers without zeros next to each other are _____

_____

_____.

8 If a number is rounded to the nearest million, it is 5 000 000.

The greatest possible value of this number is [　　　　] and the

least possible value is [　　　　].

5.2 Large numbers and rounding (2)

 Learning objective Read, write and round large numbers

 Basic questions

1 Complete each statement.

(a) Starting from the ones place to the left, the sixth place is the

_____ place and the ten billions place is the

_____ place.

(b) The least 7-digit number with 3 zeros and 4 fours is [   ] .

(c) If the hundred millions place of a number is its highest value place, it is

a [   ] -digit number.

(d) A number consists of 3 hundred-millions, 5 millions, 2 thousands and 1

one. The number is [   ] .

(e) Two hundred and six million and sixty is written in numerals as

[   ] .

(f) One billion, eight thousand and three is written in numerals as

[   ] .

(g) 25 700 890 is an [   ] -digit number. The digit 5 is in the

_____ place and it represents five _____ .

(h) Starting from the left, the first 7 in the number 7 807 321 is in the

_____ place, the second 7 is in the _____ place,
and the difference between the values they stand for is

[   ] .

(i) If a number is rounded to the nearest million, the result is 258 million.

The greatest possible value of this number is ⬚

and the least possible value is ⬚.

2 Multiple choice questions. (For each question, choose the correct answer and write the letter in the box.)

(a) In the following numbers, the number with 2 in the millions place is ⬚.

   A. 202 205 808   B. 220 208 505   C. 20 208 505   D. 2 220 208 505

(b) One hundred and one thousand, three hundred is written as ⬚.

   A. 1 001 300   B. 101 000 300   C. 101 300   D. 1 010 300

(c) Seven billion, five hundred thousand and sixty is written as ⬚.

   A. 700 050 060   B. 700 500 060

   C. 7 000 500 060   D. 40 070 050 060

(d) If a multi-digit number has only non-zero digits in its ten thousands place and ones place, then it is a ⬚ at least.

   A. 4-digit number   B. 5-digit number

   C. 6-digit number   D. 7-digit number

(e) Rounding the number 19▉324 to the nearest ten thousand, the result is 190 000. The least number that can be filled in the space is ⬚.

   A. 0   B. 4   C. 5   D. 6

3 Complete each table according to the instructions.

(a) Rounding off the number.

	To the nearest ten thousand	To the nearest million	To the nearest hundred million
769 008 000			

(b) Rounding up the number.

	To the nearest ten thousand	To the nearest million	To the nearest hundred million
5 210 178 900			

(c) Rounding down the number.

	To the nearest ten thousand	To the nearest million	To the nearest hundred million
1 094 507 260			

Challenge and extension questions

4 Think carefully and fill in the spaces with suitable numbers.

(a) For rounding off 14▨995 to get 150 000, the digit in the space can be

either _____.

(b) For rounding off 64▨995 to get 640 000, the greatest possible digit in

the space is _____.

(c) For rounding off 3674▨017 to get 36 750 000, the least possible digit

in the space is _____.

(d) For rounding up 99▇345 to get 1 000 000, the digit in the space can

be _____.

(e) For rounding down 3874▇017 to get 38 740 000, the greatest

possible digit in the space is _____.

5 Let's play a number game: I am thinking of a 6-digit number. The first three digits starting from the left are all the same. The last three digits are consecutive whole numbers, counting down to 1. The sum of the six digits equals the 2-digit number at the end of the number. What number am I thinking of?

_____

_____

_____

5.3 Four operations of numbers

Learning objective Use the order of operations, including brackets, to solve problems

Basic questions

1 Two pupils work out the answer to 8140 − 140 ÷ 4 = ☐ .

Joan's working:

8140 − 140 ÷ 4

= 8000 ÷ 4

= 2000 ☐

Mary's working:

8140 − 140 ÷ 4

= 8140 − 35

= 8105 ☐

Which method is correct? Put a ✓ for yes or a ✗ for no in each box and give your reason.

---

2 Write the order of operations in each calculation and then work it out step by step. The first one has been done for you.

(a) 75 + 25 × 6

> First multiply and then add:
>
> 75 + 25 × 6
>
> = 75 + 150
>
> = 225

(b) (75 + 25) × 6

(c) 50 × 3 + 45 ÷ 3

(d) 50 × (3 + 45) ÷ 3

3 Find the answers to these calculations.

(a) $960 - 78 \times 12 + 18$

(b) $(960 - 78) \times (12 + 18)$

(c) $(864 - 272) \div 16 + 24 \times 11$

(d) $[(864 - 272) \div 16 + 24] \times 11$

4 Multiple choice questions. (For each question, choose the correct answer and write the letter in the box.)

(a) $1700 - 1300 \div 25 \times 4 = \boxed{}$.

    A. 4   **B.** 64   **C.** 1687   **D.** 1492

(b) $(1700 - 1300) \div 25 \times 4 = \boxed{}$.

    A. 4   **B.** 32   **C.** 40   **D.** 64

(c) The sum of 60 and 20 is divided by their difference.

The number sentence is $\boxed{}$.

    A. $(20 + 60) \div (60 - 20)$

    B. $(20 + 60) \div 60 - 20$

    C. $20 + 60 \div (60 - 20)$

    D. $20 + 60 \div 60 - 20$

Challenge and extension questions

5 Fill in the boxes first. Then write number sentences with mixed operations and work out the answers.

(a) 2000 $\xrightarrow{\div\ 4}$ ☐ $\xrightarrow{-\ 323}$ ☐ $\xrightarrow{\times\ 5}$ ☐

Number sentence: _____

_____

(b) ☐ $\xrightarrow{\div\ 23}$ ☐ $\xrightarrow{-\ 76}$ ☐ $\xrightarrow{\times\ 50}$ 600

Number sentence: _____

_____

6 Write the same number in the ☐ in each sentence to make the equation true.

(a) (☐ − ☐) × 5 + ☐ ÷ ☐ = 1

(b) (☐ + ☐ − ☐) ÷ (☐ ÷ ☐) = 2

(c) ☐ ÷ ☐ + (☐ + ☐) ÷ ☐ = 3

(d) ☐ × ☐ × 2 ÷ (☐ + ☐) = 4

5.4 Properties of whole number operations (1)

Learning objective Use subtraction properties to calculate efficiently

Basic questions

1 Work these out mentally and then write the answers.

(a) 79 + 3 + 6 = ⬚

(b) 430 − 90 − 10 = ⬚

(c) 17 + 20 − 3 × 1 = ⬚

(d) 96 − 16 − 4 = ⬚

(e) 80 − 2 × 0 = ⬚

(f) 151 − (51 + 11) = ⬚

In solving the above questions, did you use any property of the subtraction operation? If so, in which questions?

_____

_____

2 Fill in the missing numbers and operation symbols to complete these calculations. The first one has been done for you.

(a) 101 − 23 − 77 = 101 − (23 + 77)

(b) 132 − (32 + 21) = ⬚ − ⬚ ◯ ⬚

(c) 277 − 11 − ⬚ = 277 − 100

(d) 919 − (⬚ ◯ 22) = ⬚ − 19 − 22

(e) _____ − _____ − _____ = a − (b ◯ c)

3 Simplify and then calculate.

(a) 800 − 246 − 154

(b) 416 − (16 + 97)

(c) 546 − (246 + 55)

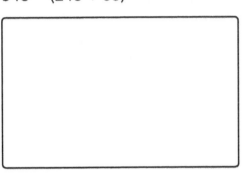

(d) 5317 − (180 + 317)

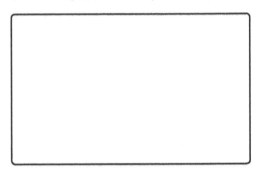

(e) 761 − 122 − 133 − 45

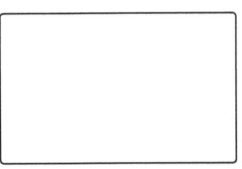

(f) 919 − 270 − 119

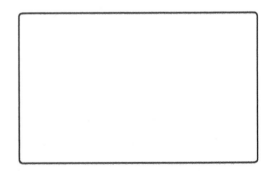

4 A bicycle factory plans to make 1600 bikes in the first quarter. It made 520 bikes in January and 480 bikes in February. How many bikes does it need to make in March? (Use two methods to find the answer.)

5 Fill in the missing numbers and operation symbols to complete these calculations.

(a) 190 − 165 + 65 = ☐ − (☐ ◯ 65)

(b) 142 − (☐ ◯ 27) = 142 − 42 + ☐

6 Simplify and then calculate.

(a) 288 − 73 − 27 + 12

(b) 3156 − (927 − 844)

(c) 483 − (216 − 183)

(d) 775 − 167 + 215 − 233

(e) (351 − 178) − (51 − 22)

(f) 6000 − 743 − 564 − 257 − 436

(g) 1000 − 1 − 2 − 3 − 4 − 5 − ... − 20

5.5 Properties of whole number operations (2)

Learning objective Use division properties to calculate efficiently

Basic questions

1 Fill in the missing numbers and operation symbols to complete these calculations.

(a) $5100 ÷ (17 × 25) = 5100 ÷ \boxed{} \bigcirc \boxed{}$

(b) $1000 ÷ 25 ÷ 4 = 1000 ÷ (\boxed{} \bigcirc \boxed{})$

(c) $128 ÷ 8 ÷ 2 = \boxed{} ÷ (8 \bigcirc 2)$

(d) $34 ÷ (\boxed{} × 2) = 34 ÷ 17 \bigcirc 2$

These answers are based on a property of division. As long as the numbers are greater than 0, we can write this as $a ÷ b ÷ c = a ÷ (b \bigcirc c)$

2 Simplify and then calculate.

(a) $7000 ÷ 8 ÷ 125$

(b) $3600 ÷ (36 × 4)$

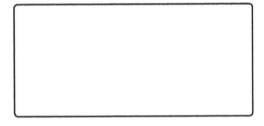

(c) $850 ÷ (17 × 2)$

(d) $360 ÷ 18 ÷ 2$

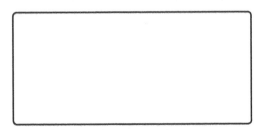

(e) $2000 \div 25 \div 4 \div 2$

(f) $5400 \div 45$

---

3 Multiple choice questions. (For each question, choose the correct answer and write the letter in the box.)

(a) $1240 \div 62 \div 2 = \boxed{}$.

   A. $1240 \div (62 \div 2)$ **B.** $1240 \div (62 \times 2)$

   C. $(1240 \div 62) \times 2$ **D.** $1240 \times (62 \times 2)$

(b) $1800 \div (25 \times 6) = \boxed{}$.

   A. $1800 \times 25 \div 6$ **B.** $1800 \div 25 \times 6$

   C. $1800 \div 25 \div 6$ **D.** $1800 \times (25 \div 6)$

(c) $9000 \div 125 \div 4 \div 2 = \boxed{}$.

   A. $9000 \div (125 \times 4 \times 2)$ **B.** $9000 \div (125 \div 4 \div 2)$

   C. $9000 \times (125 \times 4 \times 2)$ **D.** $9000 \times (125 \times 4) \div 2$

---

4 (a) 240 pupils took part in the dance performance in a school's game day. They were grouped equally into 12 teams and each team was further grouped into two sub-teams. How many pupils were there in each sub-team on average? (Use two methods to find the answer.)

(b) Lily's class plans to plant trees. The whole class is equally grouped into four teams. £240 is used to purchase saplings. Each team plants six saplings on average. How much does each sapling cost?

_____

_____

_____

Challenge and extension question

5 Calculate smartly.

(a) $540\,000 \div 125 \div 45 \div 8$

(b) $(91 \times 27 \times 76) \div (9 \times 19 \times 13)$

5.6 Properties of whole number operations (3)

Learning objective Use division properties to calculate efficiently

Basic questions

1 Complete each calculation and then fill in the spaces to make each statement correct.

$24 \div 8 = \boxed{}$

$(24 \times 10) \div (8 \times 10) = \boxed{}$

$(24 \times 3) \div (8 \times 3) = \boxed{}$

$(24 \div 4) \div (8 \div 4) = \boxed{}$

$(24 \div 8) \div (8 \div 8) = \boxed{}$

$(24 \times 100) \div (8 \times 100) = \boxed{}$

Observing the calculations above, we can find that when both the dividend

and the divisor are _____ or _____

by the same number (except zero), their _____ remains unchanged. These answers are based on a property of

_____. As long as the numbers are greater than 0, we can write this as:

$a \div b = (a \times c) \div (b \bigcirc c)$

$a \div b = (a \div c) \div (b \bigcirc c)$

133

2 Find the number sentences that have the same answer as (a). Put a ✓ or a ✗ in the box.

(a) 375 ÷ 125 = 3

(b) (375 ÷ 3) ÷ (125 ÷ 3) ☐

(c) (375 ÷ 5) ÷ (125 ÷ 25) ☐

(d) (375 × 3) ÷ (125 ÷ 3) ☐

(e) (375 × 10) ÷ (125 × 10) ☐

3 Using the above property of division calculation, fill in the spaces. Complete the statements for each division calculation by filling in the missing numbers.

(a) The quotient of two numbers is 24. If the dividend is divided by 8 and the quotient remains unchanged, the divisor should

be _____ .

(b) One number is divided by another. If the dividend is divided by 10 and the divisor is also divided by 10, the quotient

remains _____ .

(c) One number is divided by another and the quotient is 71. If both the dividend and the divisor are multiplied by 11, the quotient is ☐ .

(d) ☐ ÷ 50 = 115 ÷ 5 = 230 ÷ ☐ = 460 ÷ ☐ .

4 Choose the most appropriate method, mental or written, to find the answer to each calculation. (Check the answers to the questions marked with * using a different method.)

(a) 3200 ÷ 160 =

(b) *2820 ÷ 170 =

(c) 6320 ÷ 90 =

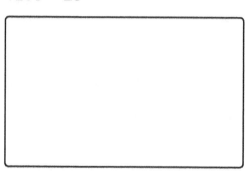

(d) *63 200 ÷ 900 =

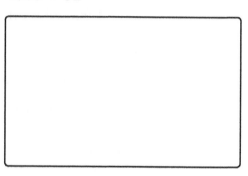

5 Calculate smartly.

(a) 1200 ÷ 25

(b) 4000 ÷ 32

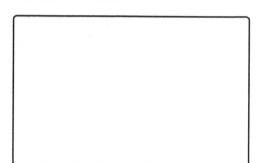

(c) 6000 ÷ 125

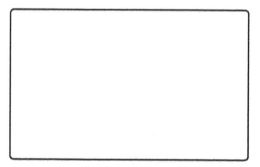

(d) 3600 ÷ 45

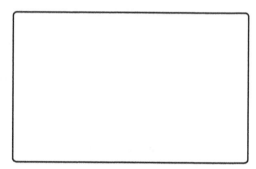

(e) 13 000 ÷ 125 ÷ 8

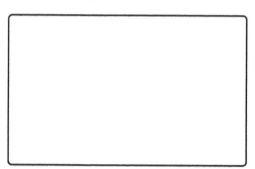

(f) 1040 ÷ 2 ÷ 52

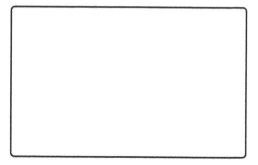

Challenge and extension questions

6 Multiple choice questions. (For each question, choose the correct answer and write the letter in the box.)

(a) The result of 4100 ÷ 700 is ☐.

    A. quotient 5 with remainder 6

    B. quotient 5 with remainder 600

    C. quotient 500 with remainder 6

    D. quotient 500 with remainder 600

(b) One number is divided by another number. If the dividend is multiplied by 2 and the divisor is divided by 2, the quotient is ☐.

    A. unchanged

    B. multiplied by 2

    C. divided by 2

    D. multiplied by 4

(c) In 120 ÷ 40, if the dividend increases by 120, then in order to keep the quotient unchanged, the divisor should ☐.

    A. increase by 120

    B. increase by 100

    C. increase by 80

    D. increase by 40

7 Think carefully and work out the answers.

(a) Two numbers are added together. If one addend is increased by 20 and the other addend is also increased by 20, what is the change to the sum?

(b) One number is subtracted from another number. If the subtrahend is decreased by 10 and the difference remains unchanged, what is the change to the minuend?

(c) The quotient of two numbers is 48. If the dividend is multiplied by 10 and the divisor is divided by 10, what is the quotient now?

(d) Two numbers are multiplied together. If one factor is multiplied by 10 and the other is divided by 10, does the product remain the same?

5.7 Roman numerals including thousands

Learning objective Read and recognise Roman numerals including thousands

Basic questions

1 Complete the table to show the value of each Roman numeric symbol in digits. One has been done for you.

Roman symbol	Value in digits
I	
V	
X	
L	50
C	
D	
M	

2 Write the following Roman numerals in digits.

XIII = _____   LVI = _____

XCV = _____   CIV = _____

CD = _____   DC = _____

MCM = _____   MMVII = _____

3 Can you recognise years written in Roman numerals? Complete the table below. The first one has been done for you.

Roman numeral	MM	MD	MDLII	MCM	MCMXCV	MMXVI
Year	2000					

Challenge and extension question

4 Write the following years in Roman numerals. The first one has been done for you.

Year	2010	2011	2012	2013	2014	2015	2020	2100
Roman numeral	MMX							

5.8 Solving problems in statistics

Learning objective Complete, read and interpret data presented in different ways

Basic questions

1 The pictogram below shows the numbers of Year 5 pupils participating in different school clubs. Use this information to find the answers to the questions.

△ △ △ △ △	△ △ △ △	△ △ △	△ △ △ △
Choir	**Science**	**Dancing**	**ICT**

Each △ stands for 5 pupils.
Which school club has the greatest number of participating pupils?

_____

Which school clubs have the same number of participating pupils?

_____

How many more pupils are there joining choir than dancing? _____
From the pictogram, can you tell the total number of pupils participating in different school clubs? Why or why not?

_____

_____

_____

2 Read the following table carefully and then work out the answers.

Results of standing long jump test (Year 5)

	Linda	Alvin	Bob	Peter	May
Result (cm)	242	256	261	228	219
Place					

(a) What is the difference between the longest jump and the shortest jump?

_____

(b) According to the results, write each participating pupil's place in the table.

_____

_____

3 The bar chart shows the production of cars each quarter for a year. Use this information to find the answers to the questions.

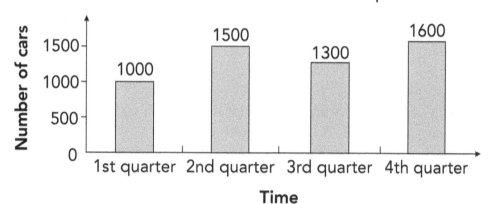

(a) The total output of the manufacturer is _____ cars in the year.

(b) Construct a line graph using the data represented in the bar chart.

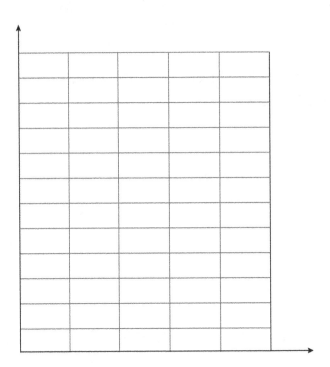

(c) According to the line graph, which of the following is true?
Put a ✓ for true or a ✗ for false in each box.

(i) The production was stable over the four quarters. ☐

(ii) The production was steadily improving over the four quarters. ☐

(iii) The production was steadily declining over the four quarters. ☐

(iv) The production fluctuated over the four quarters. ☐

4 The table presents a summary of the water bill of Ivan's family for the second quarter of 2017 (the unit prices have been rounded to the nearest penny). Read it carefully and answer the questions.

	Usage (m³)	Unit price	Charge
Fresh water used	79	120p	
Used water disposal	73	227p	
		Total charge	

(a) Use the data shown in the table to work out the charges and complete the table.

(b) How much water did the family use? Express it in both cubic metres and litres. _____

(c) Excluding other standing charges, how much did the family need to pay for their water bill for this period?

_____

(d) Get a recent water bill from your home and make a table like the one above. Are there other standing charges?

Chapter 5 test

1 Work these out mentally and then write the answers.

(a) $20 \times 50 - 80 =$ ⬚

(b) $230 + 90 + 70 =$ ⬚

(c) $900 - 99 + 1 =$ ⬚

(d) $96 \div 12 \times 12 =$ ⬚

(e) $45 \times 7 + 3 =$ ⬚

(f) $75 - 75 \div 5 =$ ⬚

(g) $770 \times 8 \div 70 =$ ⬚

(h) $16 \times 5 \times 2 =$ ⬚

(i) $32 + 8 \times 2 =$ ⬚

(j) $\frac{8}{15} - \frac{11}{30} + \frac{7}{15} =$ ⬚

(k) $6400 \div 200 =$ ⬚

(l) $10\,\text{kg}\,20\,\text{g} - 920\,\text{g} =$ ⬚ g

2 Use the column method to calculate.

(a) $56\,400 \div 700 =$

(b) $180 \times 209 =$

(c) $71\,200 \div 2300 =$

3 Work these out step by step. (Calculate smartly when possible.)

(a) $412 \div (607 - 36 \times 14)$

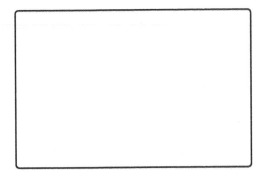

(b) $72 \times 36 + 18 \times 36$

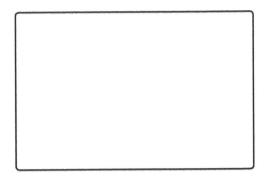

(c) $80 \times [(325 + 7) - 304]$

(d) $250 \times 8 - 10\,000 \div 8$

(e) $(606 - 330) \times (116 + 434 \div 14)$

(f) $36 \times [(99 - 21) \times 99 + 108]$

(g) $3\frac{6}{7} + \frac{11}{35} - \frac{33}{35}$

(h) $\frac{1}{100} + \frac{99}{100} - \frac{7}{9}$

(i) $3 \times \frac{4}{7} - 5 \times \frac{3}{14} + 7 \times \frac{5}{28}$

4 Fill in the spaces to make each statement correct.

(a) 1000 _____ is a million. ☐ millions is a billion.

(b) 120 456 consists of ☐ thousands, ☐ hundreds, ☐

tens and ☐ ones.

(c) Put the numbers 1, $\frac{5}{6}$, $\frac{4}{7}$ and $\frac{5}{7}$ in order, starting from the greatest.

The second number is ☐ .

(d) Pour 3 litres of vegetable oil into 450 ml bottles. ☐ bottles will be

filled up and ☐ bottles are needed in total.

(e) Rounding the number 555 500 to the nearest ten thousand,

it is ☐ .

(f) When rounding down the number 10 700 342 to the nearest ten

thousand, it is ☐ . When rounding up the number to

the nearest hundred thousand, it is ☐ .

(g) A 1 metre-long string was divided into 100 equal parts. Six parts of it

are ☐ metres long.

(h) 4800 ÷ 160 = 480 ÷ ☐ = ☐ ÷ 80 = ☐ ÷

☐

(i) In 6400 ÷ 3200 = 2, if the divisor is decreased to 64 and the quotient

remains unchanged, the dividend should be ☐ .

(j) When one number is divided by another, the quotient is 17 and the
remainder is 100. If both the dividend and the divisor are multiplied by

10, then the quotient is ☐ and the remainder is ☐ .

(k) After a number is rounded down to the nearest ten thousand, it is 400

thousand. The greatest value of the number could be ☐ .

(l)   The quotient of two numbers is 45. If the dividend remains unchanged and the divisor is multiplied by 3, then the quotient is ☐.

(m)   One number is divided by another and the quotient is 6. If both the dividend and the divisor are multiplied by 74, then the quotient is ☐.

(n)   Both the dividend and the divisor are divided by 100 and the quotient is 4 with the reminder 2. Given that the original divisor is 700, the original dividend is ☐.

5 True or false? (Put a ✓ for true and a ✗ for false in each box.)

(a) $27 \div 9 = (27 \times 3) \div (9 \times 3)$ ☐

(b) $54 \div 6 = (54 \div 2) \div (6 \times 2)$ ☐

(c)   In $460 \div 50$, the quotient is 9 and the remainder is 1. ☐

(d) $560 \div 16 = (560 + 8) \div (16 + 8)$ ☐

(e)   The year 2017 in Roman numerals is MXVII. ☐

6 Multiple choice questions. (For each question, choose the correct answer and write the letter in the box.)

(a)   The number sentence equal to $72 \div 24$ is ☐.

   A. $(72 \times 2) \div (24 \div 2)$ **B.** $(72 \div 24) \times (24 \times 4)$

   C. $72 \div 6 \div 2$ **D.** $(72 \times 3) \div (24 \times 3)$

(b)   From $13 \div 4 = 3 \text{ r } 1$, we can get $1300 \div 400 = $ ☐.

   A. 3 r 10   **B.** 30 r 10

   C. 3 r 100   **D.** 300 r 100

(c)   The number sentence not equal to 101×199 is ☐.

   A. $(200 + 1) \times 101$ **B.** $199 \times (100 + 1)$

   C. $199 \times 100 + 199$ **D.** $(200 - 1) \times 101$

(d) After dissolving 1 gram of sugar in 99 grams of water, the mass of the sugar water solution is ☐ the mass of the sugar.

    A. (99 + 1) ÷ 1 times   **B.** 99 ÷ 1 times

    C. 1 ÷ (1 + 99) times   **D.** 1 ÷ 99 times

(e) MDCXI in Roman numerals is ☐ in digits.

    A. 1609   **B.** 1611   **C.** 1409   **D.** 1561

7 Write the number sentences and work out the answers.

(a) Number B is 150, which is half of Number A. What is the sum of Number A and Number B?

(b) The difference between 48 and 16 is multiplied by 25 and then divided by 8. What is the quotient?

(c) The difference between a number and 12 is divided by 125, and the quotient is 8. What is the number?

8 Solve these problems.

(a) A school planned to donate 780 books to schools in a disaster-hit area. The actual number of books it donated was 66 more than 3 times the number of books it planned. How many books did the school donate?

(b) A car travelled from City A to City C via City B in 4 hours. (See the diagram below.)

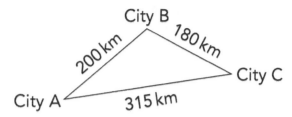

(i) What was the average speed per hour the car travelled?

_____

(ii) When going back from City C straight to City A, the car travelled 10 km more per hour. How many hours did it take the car to reach

City A? _____

_____

_____

_____

(c) A school bought 12 basketballs, which is half the number of volleyballs it bought. The number of footballs the school bought is 7 more than 3 times the total number of basketballs and volleyballs. How many footballs did the school buy?

(d) An art studio bought 244 boxes of red pens and green pens. The number of boxes of green pens is 12 boxes fewer than 7 times the number of boxes of red pens. How many boxes each of red pens and green pens did the studio buy?

(e) The table shows the number of pupils in a school whose birthdays fall in different quarters. Look at the table carefully and answer the questions.

	1st quarter	2nd quarter	3rd quarter	4th quarter
Number of pupils	130	180	95	155

(i) There are [   ] pupils in total in the school.

(ii) Construct a line graph using the data represented in the table.

(iii) From the table or the line graph, can you tell the exact number of pupils in each year group? Can you give an estimate? Explain.

_____

_____

_____

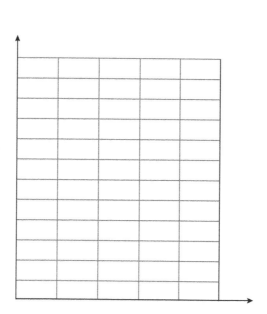

Notes

Notes

Notes

Notes

Notes